PENGUIN BOOKS

AROUND THE WORLD IN 68 DAYS

Koh Buck Song (family name: Koh) is a Singaporean brand adviser and the author and editor of more than 30 books, including one on Singapore's country brand—*Brand Singapore: Nation Branding In A World Disrupted By Covid-19* (third edition, 2021, with a Chinese translation published in China, 2012). He has advised foreign governments on building their nation brands, and has spoken extensively on country branding internationally, including in London, Oxford, Cambridge (Massachusetts), Chicago (Illinois), Phuentsholing (Bhutan), Tahiti, Tokyo, Shanghai and Melbourne.

An avid traveller, he has visited more than 80 countries and territories, from Antarctica to Zimbabwe. An artist of *haiga*, a modern interpretation of the 16th-century Japanese art form of ink sketches combined with *haiku* poetry, he has held several exhibitions in Singapore and Laos. This book features a *haiga* artwork for each of the 13 countries visited. He read English at Cambridge University in the UK and has a master's in public administration from Harvard Kennedy School, USA.

As an adjunct editor at Singapore's Centre for Liveable Cities, he has authored and edited reports on the annual World Cities Summit since 2014. Previously, he was a columnist and political and arts journalist with Singapore's main English daily *The Straits Times*, and head of global media relations and strategic planning at the Singapore Economic Development Board. As head of public affairs (Southeast Asia) with the communications consultancy Hill & Knowlton, he advised the Singapore government on many aspects of urban development, including the global launch of Gardens by the Bay and the National Gallery Singapore.

RESPONSES TO AROUND THE WORLD IN 68 DAYS
FROM ALL 13 COUNTRIES, IN THE ORDER OF TRAVEL

'With a poet's eye for detail, a spiritual outlook lightly worn and a gift for capturing the unique brand of each locale visited, Koh Buck Song's sensitive snapshots of his fascinating and far-flung escapades offer a timely dose of armchair travel in our pandemic times.'

—Umej Bhatia
Singapore's Ambassador to the United Arab Emirates
(2013–2016) and Consul-General in Dubai (2011–2012)

'Koh, the traveller who turned into a storyteller, left me speechless after a journey to 13 countries, living his inspirational stories and seeing my own country anew with his words.'

—Mohammed Al Ardhi
Executive Chairman, Investcorp,
and Chairman, Sohar International, Muscat, Oman

'Koh gives deep insight into the paradox of Madagascar, a resource-rich country in a permanent state of survival.'

—Tahiry Sojarison
Ministry of Foreign Affairs, Madagascar

'Koh brings out the best of Kenya, its people, flora and fauna. Travelling to 13 countries in 68 days is a dream for many, myself included—*ahsante sana* for including my beautifully unique Kenya on your itinerary.'

—Dr Wanja Njuguna
Journalist and academic, Lake Naivasha, Kenya

'Koh's "momentary glimpse" of the "cultural salad bowl" of Spain offers, in a few pages, a deep immersion into the character of Bilbao. Like the Guggenheim's architect Frank Gehry, Koh connects old with new, global with local, with a perspective that blends the fresh sight of the tourist with that of the global expert in place branding.'

—César Arjona
ESADE Law School, Barcelona, Spain

'Koh's work, fostering a love for the different cultures of the world, ignites a curiosity and interest that will lead readers to witness Colombia's history, unprecedented evolution and commitment to peace and legality.'

—Manuel Solano
Ambassador of Colombia in Singapore

'Explaining the meaning of Costa Rica's heartfelt saying pura vida is quite a challenge already. Thanks to Koh's understanding of our diverse world of colours, flavours and life, he depicts it masterfully, elaborating on the essentials that describe our approach to wellbeing and happiness.'

—Daniel Valverde
Essential COSTA RICA Country
Brand Director, Costa Rica

'Written with accuracy and a beautiful new perspective, Koh takes us on a marvellous chronicle of his tireless journey to discover the natural charms, culture and soul of some magical towns.'

—Fredesman Turro Gonzalez
Ambassador of Cuba in Singapore

'In 68 days, Koh transports you around the world and, at the same time, teaches about history, culture, and how both aspects define much of our present and shape our future.'

—Martin Torrijos
President of Panama (2004–2009)

'Trips are not measured in kilometres, but by streets, landscapes, faces, aromas—by the unique moments that make their home in the soul of the traveller so they are never lost. With Koh's book, you will have a part of Mexico in your heart.'

—Agustín García-López Loaeza
Ambassador of Mexico in Singapore

'Koh's book, the result of incredible work, has captured the essence of significant parts of Guatemalan society.'

—Luz Mendez
President, National Union of Guatemalan Women, Guatemala

'Koh is a 21st century Matsuo Basho. His ichigo ichie spirit, that infuses his haiga art, will keep stirring our yearning to travel.'

—Misako Ito
Japan Foundation, Tokyo, Japan

'I admire Koh's keen insights into the daily, cultural rhythms of society often missed by locals. This book will awaken you to a new appreciation of the people, places and ways of life in your hometown or your next journey.'

—Gina Chang
H I Publishers, Taipei, Taiwan

Around the World in 68 Days

Observations of Life from a Journey Across 13 Countries

Koh Buck Song

PENGUIN BOOKS
An imprint of Penguin Random House

PENGUIN BOOKS

USA | Canada | UK | Ireland | Australia
New Zealand | India | South Africa | China | Southeast Asia

Penguin Books is part of the Penguin Random House group of companies whose addresses can be found at global.penguinrandomhouse.com

Published by Penguin Random House SEA Pte Ltd
9, Changi South Street 3, Level 08-01,
Singapore 486361

Penguin Random House SEA

First published in Penguin Books by Penguin Random House SEA 2021

10 9 8 7 6 5 4 3 2 1

The views and opinions expressed in this book are the author's own and the facts are as reported by him which have been verified to the extent possible, and the publishers are not in any way liable for the same.

ISBN 9789814882231

Typeset in Adobe Garamond Pro by Manipal Technologies Limited, Manipal

www.penguin.sg

For
Dora

the love of my life
and travel planner/organizer extraordinaire

The author would also like to express deep appreciation and gratitude to Umej Bhatia, Mohammed Al Ardhi, Tahiry Sojarison, Wanja Njuguna, César Arjona, Manuel Solano, Daniel Valverde, Fredesman Turro Gonzalez, Martin Torrijos, Agustín García-López Loaeza, Luz Mendez, Misako Ito, Gina Chang and Nora Nazerene Abu Bakar.

Contents

Prologue

Ichigo ichie and the wonder of travel

This book is a gathering of observations, souvenirs of insight recollected in tranquillity, drawn from a trip that my wife Dora and I completed in sixty-eight days in the late spring and early summer of 2018, or the year 1 BC (Before Covid-19). The global pandemic that emerged at the end of 2019 has upended life on planet Earth. The disease is still in the process of transforming human existence and will certainly also reset the world of travel. The 'BC era' was a time that now seems so far away, a time when the global economy, powered by free movement of people and goods and services, was truly thriving and buzzing. It was a time when people were sometimes too preoccupied to give thanks for the joys of life, too busy to dream. Not only would some not see the roses, let alone smell them; but they might ignore you even if you pointed out those roses.

But others sought to live in the moment, and some knew there was a lot more world to experience. This was humanity's golden age of travel; more people were seeing the world than ever before. 2018 saw a record 1.4 billion international tourist arrivals, according to the World Tourism Organization, a rise of 6 per cent over 2017. People were beginning to realize that meaningful life experiences—like those one enjoys as a visitor to a foreign land—ultimately matter more than material possessions. As the YOLO ('you only live once') spirit caught on, more people figured to themselves: If life is precious and short, what better way is there to live than to see and enjoy as much of it as possible? Covid-19 has most probably taken this to a whole new level, the full measure of which might be too early to grasp.

Frequent flyer programmes were constantly adding new benefits in an unrelenting bid to outdo their rivals. The airline industry was expanding exponentially. As business class seats reclined flatter and flatter, tourism growth charts curved up more and more sharply. Budget airlines—with their competitive, sometimes unbelievably low, fares—were touting slogans like 'now everyone can fly'. For the optimists amongst us, the world was coming together in that fabled 'global village', all of us learning more about our fellow men, becoming that little bit more empathetic, and thus, becoming closer as fellow residents and stewards of this fragile planet. With the pandemic, this 'global village' has gone virtual, connecting in new ways all of us, cocooned in our 'little huts'.

For me, setting off on this round-the-world voyage in 2018 was a dream come true. At that time, I knew it would be a precious experience, no matter how it turned out. Now, it has become even more so in a Covid-19 era, when, amidst all the unpredictability, one thing is certain—hardly anything will return to being exactly as it was before.

Within the pages of this book, there are no expert analyses, merely momentary glimpses into thirteen societies, in a time before Covid-19, by someone who simply had the good fortune to visit these places, to have an opportunity to feel their essence, if only for a while.

Our itinerary covered thirteen territories in this order: the United Arab Emirates (UAE), Oman, Madagascar, Kenya, Spain, Colombia, Costa Rica, Cuba, Panama, Mexico, Guatemala, Japan and Taiwan. In this journey, a few key aspects coalesced: the wonder of travel in general, and in particular, that special category of a continuous trip around the globe; the magic of each moment of discovery and insight on the road; and the joy of glimpsing the essence of a range of societies that are so different from each other and yet have so much in common. To us, it was worth investing a chunk of our hard-earned life savings from decades of work to pay for this trip, for what is life if not to be savoured?

Starting in Singapore, the route took in terrain of varied contrasts, from skyscrapers shooting out of barren, empty desert in the United Arab Emirates to the lush 'botanical Galapagos' of Madagascar, replete with unique plants and wildlife. Despite human intervention and damage, the endless marvels of nature abound everywhere, from the 'big five' animals of Kenya to the smallest creatures of Costa Rica.

In the massive gallery of this globe, brushstrokes of the human story have been painted on many, many canvases. Each trip is like a peek into a kaleidoscope, and travel can make possible only brief glances through each lens. Narratives of how man has always striven to survive and thrive vary according to vastly different environments. Colonialism's legacy can be examined everywhere, from a former imperial power like Spain, still a formidable cultural force, to its former colonies such as Colombia and Guatemala, still recovering from the political, economic and cultural invasions and impositions of eras past, or, in the case of Cuba, still navigating a way out of isolation towards some semblance of regional and global connection.

Just as fascinating are the case studies of the enduring legacies of ancient civilizations, from the Maya and Aztecs in Mexico to the Ryukyu people in Japan. And always intriguing is the condition of small states seeking their place on the world stage, whether, as in Oman, finessing a delicate balance between preserving tradition and pursuing modernity; or, as in Panama, being buffeted literally by the waves from two oceans and figuratively from global competition; or, as in Taiwan, yearning to emerge from the shadows of its past. Covid-19 has called a timeout on all these human struggles. But the indomitable human spirit will surely rise again, and resume these yearnings for survival, success and support from other nations.

Within the pages of this book, there are no expert analyses, merely momentary glimpses into thirteen societies, in a time before Covid-19, by someone who simply had the good fortune to visit these places, to have an opportunity to feel their essence, if only for a while, through all the five senses: To see the vastness and variety of the wilderness, from a cheeky cloud dancing on the edge of a mountain peak to the tiniest, hardly visible, tree frog. To hear the unbridled joy of a hummingbird's chirp and the possible jealousy in a lioness' growl. To smell the spices and array of savouries in a souq, and the accumulated natural fertilizers of the savannah. To touch the outstretched reaches of humanity, from the metallic-smooth terrace balcony of the world's tallest skyscraper at sunrise to the rough, weathered stone walls of an ancient monument of sun worship. To taste the Earth's boundless nutrition and nectar, from the sashimi of southernmost islands to the juice of a baobab fruit.

On top of all this, also to tune the sixth sense to resonate with the true spirit of a people, navigating a course between materialism and spirituality, between aspiration and anguish.

For some time, I had been living life with the acute awareness that most experiences in life are unique, and therefore special, because they will never be repeated. Sometimes, it can be something as commonplace, and seemingly commercial, as when a restaurant you had been wanting to try closes down before you could get round to it. Once, I drove a friend visiting from overseas to a favourite Peranakan restaurant off Orchard Road in Singapore, only to find, when we pulled up towards the driveway, that the whole hotel building had been demolished and all that remained was a construction site around a hole in the ground.

The pandemic has brought home the wisdom of ichigo ichie even more starkly . . . If ichigo ichie should already transform everyday life, then its power becomes that much stronger when one is travelling.

The deeper spiritual truth is that nothing should ever be taken for granted—not a random stranger you've just met while asking for directions as a tourist or while networking at a conference, and not even the smallest moth or worm you might spot in a garden. In the realm of personal ties, this idea of making the most of opportunities in life is driven home most clearly—and often, painfully—when a loved one departs, either succumbing unexpectedly to mortality or when the relationship breaks down irreparably. Many a word goes unuttered, a sentiment unexpressed, because we think there will be another occasion, another tomorrow. And then, in the blink of an eye, that moment will have passed—forever.

This realization was with me long before I came to learn of the marvellous Japanese concept of ichigo ichie, based on the notion that life becomes that much more meaningful when you are always ready to make the effort to treasure any encounter of potential in life. What makes a seemingly mundane experience into something much more usually comes from little details. As a bonus, a coincidence relating to the fortuitous matching of words or numbers or colours or some other aspect makes a juxtaposition even more significant, like meeting a friend for lunch and discovering that you're both dressed in clothes of the same hue, by chance the 'corporate colours' of the primary school you both went to. Being observant and perceptive enough to see such details of significance can reveal deep insights that give added reason to cherish that moment. Now, the pandemic has brought home the wisdom of ichigo ichie even more starkly.

Ichigo ichie is, in some ways, more profound than the better-known phrase carpe diem (Latin for 'pluck' or 'seize' the day), attributed to the ancient Roman poet Horace, meaning to enjoy the pleasures of the moment while minimizing concerns for the future. By contrast, ichigo ichie can be said to advise quite the opposite—to be more fully aware of the future so as to appreciate the present even better. If ichigo ichie should already transform everyday life, then its power becomes that much stronger when one is travelling. The greatest motivation, and reward, for those who love to see the world is the sense that everything one sees in a foreign country is fresh, interesting and worth remembering. Visiting a new place is already a treat. Going around the world takes the awareness of ichigo ichie to a whole new level.

This journey traversing the planet was something I could not have imagined before it happened. The trip was made possible only with the incredible resourcefulness and planning excellence of my dear wife Dora. I could only stand back and admire her ability to assemble such a complicated itinerary across thirteen countries, and to organize everything by booking directly with ground agencies, hotels, restaurants and other travel providers. Her inspired curation of such a diverse and fulfilling range of experiences was exceptional, including some of the best of nature, culture, heritage accommodations and local flavours. This valuable capability of hers I had already noted from our earlier travels. But with this trip, she outdid herself big time. Now, looking back through the lens of a Covid 19 world, this was truly a journey of a lifetime.

The literary tradition of circumnavigating the world

This book pays tribute to a beloved literary tradition. Its title obviously alludes to the adventure novel *Around The World In Eighty Days* by the French writer Jules Verne (1828–1905), first published in French in 1872, and in English translation a year later. The plot centres on an Englishman, Phileas Fogg, who makes a trip with his French valet Passepartout across the globe, so as to try to win a £20,000 wager made after an impetuous argument in a London café. The bet was that eighty days would be enough time for such a journey, given new 'disruptive innovation' developments in transportation at that time such as the completion of the First Transcontinental Railroad in the United States of America (1869), the linking of the railways across the Indian subcontinent (1870) and the opening of the Suez Canal in Egypt (1869).

This classic celebration of travel and imagination inspired in readers dreams of themselves planning and embarking on a trip that would previously have been possible only for the most hardy and ambitious adventurers. Verne's story has been retold in a few movies, including one in 1956 starring David Niven as Fogg and a 2004 Disney live-action film with Jackie Chan as Passepartout. On TV, the adaptations include a 1989 miniseries featuring Pierce Brosnan as Fogg.

The fundamental connection between Verne's novel and this book is in the excitement of circumnavigating the whole planet, which evokes universal fascination, even when there are still people who believe the Earth is flat. There are many places one could visit separately around the world, yet there is something special about setting out on a trip in one direction and to keep going around the entire span of the planet until you finally reach home again.

Verne's adventure included ten stops across eight countries—Paris, Brindisi, Suez, Bombay, Calcutta, Singapore, Hong Kong, Yokohama, San Francisco and New York—before returning to London. His feats of imagination were prodigious in blending fiction with facts crafted together from his own knowledge and research.

For example, in chapter seventeen, his description of a two-hour stopover—a kind of layover city-tour—in Singapore for his ship's refuelling contains details that quite accurately evoke what life would then have been like in this nineteenth-century British colony: 'Pepper plants replaced the prickly hedges of European fields; sago-bushes, large ferns with gorgeous branches, varied the aspect of this tropical clime; while nutmeg-trees in full foliage filled the air with a penetrating perfume.

Agile and grinning bands of monkeys skipped about in the trees, nor were tigers wanting in the jungles.'

The Penguin Classics 1873 English edition of Jules Verne's adventure novel *Around The World In Eighty Days.*

Much of the impetus in Verne's story is driven by the protagonist's race against time. By contrast, in this book, there is no race except to reach an airport in time for a flight. And even then, the only time there was anything approaching a true race was in Antananarivo, capital of Madagascar, on our way to catch an early afternoon flight to Nairobi, Kenya. Our guide, Eperlon, had arranged to set off from Andasibe National Park after a very early breakfast so as to give enough buffer time, slightly more than the industry average. Although we ran into a typical midday traffic jam on one of the main roads leading through the city to the airport, it seemed at first as if everything would be fine. But then, the jam seemed worse than normal, so our guide instructed our driver, Lova, to try another route. When this smaller secondary road was no smoother, they opted for a third axis.

Our guide just about kept his composure, even though the stress he was feeling was clearly visible, from his facial expressions viewed sidelong from the back seat and his immoderately frequent glances at his watch. His anxiety was audible too—from his conversation in the local language of Malagasy with our driver, we could tell that they were getting increasingly worried. More than once, our driver had to make a three-point turn in the middle of a tight alley, narrowly avoiding hitting other vehicles as well as tourists and locals on foot, so as to abandon the route we had been on and turn to head for an alternative way. This went on until we finally arrived at the airport, with only minutes to spare. We had come that close to sabotaging the rest of the trip.

Such a long journey is fraught with a few inevitable apprehensions, which one who embarks on such a voyage must suppress, and keep subdued, right till the very last leg of the journey. Chief amongst them is some unforeseen delay or obstacle that could jeopardize the entire enterprise. For this trip, the pressure was on from the first check-in in Singapore to ensure that no flight was missed. This is because the conditions of our air tickets stipulated that any no-show for any leg of an air itinerary would mean the cancellation of all the remaining portions of the round-the-world flight bookings. Apparently, the rationale for this regulation is to impose a penalty to discourage no-shows.

We had to keep going. Like Jules Verne, we needed stamina and perseverance.

So, we had to keep going. Like Jules Verne, we needed stamina and perseverance. The main Star Alliance flight itinerary stipulated no backtracking and allowed for connections only with partner carriers. However, we got around this limitation by weaving in side-trips made on other domestic airlines, so as to cover more destinations. This was how the resulting itinerary provided for explorations of a wider range of fascinating terrain and intriguing cultures.

The reasons are very different, but the same need to keep to the itinerary is there in *Around The World In Eighty Days,* as Verne writes: 'A single delay would suffice to fatally break the chain of communication; should Phileas Fogg once miss, even by an hour, a steamer, he would have to wait for the next, and that would irrevocably render his attempt (to complete this trip around the world) vain.' To make it through the whole itinerary, reserves of patience, flexibility, resourcefulness and more than a touch of Zen were called upon. Indeed, what an adventure it all turned out to be.

Seeking the quintessence of nations

This book might be shelved in the travel section of a bookshop or library, but it is not about how to get there, what to see or where to eat. With the disruptions of Covid-19, this book is now, in a sense, a memoir and celebration of travel in an era that is no more. But more importantly, this is a collection of ruminations about the quintessence of nations. One connecting thread across the thirteen countries visited is my effort to perceive the true nature of each country, which is also its brand essence. This usually revolves around a few characteristics that are at the heart of a nation, society and people. Often, national icons like prominent buildings are important symbols, such as the Burj Khalifa of Dubai or the Taipei 101 tower in the Taiwanese capital. Sometimes, national metaphors are a vital part of the discussion, like whether the lemur or the zebu (an unusual species of cattle) would be a better representative of Madagascar.

Wearing a traditional lamba shawl, presented by the government of Madagascar, in the capital Antananarivo.

Place branding is a field of work that is about understanding, conceptualizing and promoting the key attributes of a place—from a specific location to an entire country—so as to contribute to the overall objective of raising the quality of life through building national identity and spurring economic growth. To put my place branding observations together from this journey, I drew from my work in the space of place branding across three decades. My work in this area, including advising foreign governments on how to build their country brands, had also given me the chance to make a few special trips, including to Bhutan, Tahiti and London. I have also had the chance, on my own trips, to speak on the subject in places including Shanghai, Melbourne and Oxford.

And so, place branding was part of the programme on this sixty-eight-day trip. In Madagascar and Costa Rica, we met place branding practitioners whom I had met previously at an international conference in London. It was a special bonus to hear from their professional and personal perspectives. Looking at a country through the lens of place branding opens up intriguing facets for further consideration. The following chapters, one for each country, distil the observations and musings collected along the way. Each insight is a tiny fragment of the mosaic of awesome variety that is human life on this planet we all share.

Mexico
Guatemala
Cuba
Costa Rica
Panama
Colombia
Spain
United Arab Emirates
Oman
Kenya
Madagascar
Singapore
Taiwan
Japan

United Arab Emirates

Heights of human aspiration

The haiku poem and ink sketch in this haiga artwork capture the primary country brand attribute of the United Arab Emirates: to always reach for the superlative. The impossible-to-ignore national initiatives range from building the world's tallest skyscraper to hosting the World Expo 2020. While the UAE's openness to connecting with other cultures is like a beacon in the Muslim world, what keeps its ambition level-headed are the Islamic virtues of humility, perspective and a willingness to learn.

Note:

Haiga is a traditional art form that dates back to sixteenth century Japan. It combines a visual artwork, usually an ink sketch, with a haiku poem of three lines, of five-seven-five syllables, usually making a simple, yet often profound, observation about everyday life. I became fascinated with haiga when I came to study it in the course of a trip in 2015 to Japan (to Tokyo, Kyoto, Yamanashi prefecture where Mount Fuji is, and Okinawa), at the invitation of the Japan Foundation as a cultural leader of Singapore. Since then, I have held several haiga exhibitions in Singapore (including at the Super Japan Festival 2016 organized by Singapore's national arts centre, the Esplanade) and in Vientiane, Laos, in 2017. Experimental modifications that have been made to the haiga form include adding a title, the use of rhyme in the haiku's first and third lines, and sometimes Singlish words.

UAE
Greater heights
where there's higher reach,
embracing the world is for
ourselves to teach
許木松
2020

Top of the world: The UAE aspires to lead the way in many aspects, and succeeds in a few.

Dubai: Begin with the top in mind

This journey begins right at the very top. Leaving Singapore, we headed to Dubai, the city that hosts (at the time of writing) the world's tallest building, the Burj Khalifa. It is a title that the 160-storey tower has held for more than a decade, since its topping-out in 2009. There is no building anywhere else in the world closer to the clouds. With Covid-19 now delaying plans for any new skyscrapers, the Burj Khalifa looks like it will certainly hold this claim for some time to come.

The summit of human achievement in engineering was an appropriate first stop for our voyage, for more reasons than one. For one thing, getting a bird's eye view of any city, whether from a skyscraper or an aircraft, always sets the scene with the best kind of perspective. It offers an overview of a place, as well as a good glimpse into the distance. This gives a sense of the space one is about to explore in greater depth. There is an African proverb that if you want to go fast, go alone, but if you want to go far, go together. I would add this: To go far well, it is good to see far before setting off.

If you want to go fast, go alone, but if you want to go far, go together . . . To go far well, it is good to see far before setting off.

Reaching a high point, literally, in a new city is something many visitors have high on their to-do lists. It is from there that one can get the best view of one's newfound surroundings. It is from there that one can identify major landmarks and their interconnections, and so, see how the city sits amidst the natural landscape that it is nestled in. It is like looking at a 'map' of the city in real life. With this overview in mind, all subsequent explorations are enhanced, because you can then relate pieces of tourism real estate to the overall geography, and also, how they relate to each other in terms of the mental map gained from that initial overview. What is more, for our trip, for eventual closure towards the end, it also set up a fitting symmetry, from one skyscraper to another, for the highlight of the final destination of our trip—the Taipei 101 tower in Taiwan.

For sure, to have a clear focus and direction sets the path to success in any human endeavour. 'Begin with the end in mind,' advises Stephen

Covey in his 1989 book *7 Habits Of Highly Effective People*. For the UAE, it is as if 'begin with the top in mind' is the national mantra for its approach to all its major initiatives, including projecting the skyline

Meeting of modern and traditional: The Burj Khalifa, seen here from a skyscraper balcony terrace late at night, frames a futuristic scene of the skyline and city lights, while the interlacing of expressway viaducts appears to evoke the most ancient of Islamic designs.

of Dubai, as a symbol to the world of what the country stands for. In putting together its city infrastructure, Dubai's aim has always been to reach for the highest point possible, and then to use such achieved heights to inspire other, further aspirations for the rest of the country.

To appreciate the brand essence of Dubai and the UAE, it was appropriate to begin with examining what had gone into creating this position of leadership amongst the countries of the world, and also the consequent effect of such an approach on the people and society. For this, the UAE presents an outstanding case study.

The idea had occurred to me before that some cities that lack major sights might look to build an observation tower, so as to add to the things that visitors can do. This is not in any way a criticism of these cities. It is just an aspect of reality of what some cities do to make themselves even more attractive to visitors, investors and immigrants. In this group of cities, Dubai is at the top of the league.

Conversely, for a city that has a 'natural' heritage sight, such as the pyramids of Giza in Cairo, Egypt, it is probably unnecessary to add

Ancient grandeur: Dubai's Wafi Mall, designed in Egyptian style, celebrates pinnacles of achievement by human civilizations, which is what the UAE aspires towards in modern times.

something like a lookout tower, since history has bequeathed to the city such a wondrous tourist attraction. Perhaps, then, there is something more to the fact that the Wafi Mall, one of Dubai's best-known shopping centres, is designed and decorated in ancient Egyptian style, reminiscent of the temple complex of Karnak. This might well at least be subconsciously congruent with a key notion in the psyche of the UAE—to look to the grandeur and scale of the major ancient civilizations as a benchmark or inspiration for what the UAE could achieve in modern times. Indeed, the Great Pyramid of Giza was, in its time, the world's tallest building, like the Burj Khalifa is today.

We had planned our visit to the Burj Khalifa at the crack of dawn, to beat the crowds. Our hotel was near enough, just a short taxi ride away. When we got to the building, it took a while to find the lobby for the elevator up. Unfortunately, there was no chance of a brilliant sunrise

Through a glass, darkly: On this typical morning, car fumes, desert haze and clouds on the horizon meant that the dawning sun broke through only when it was already at a height in the sky, in this view of Dubai from the 125th-level observation deck of the Burj Khalifa.

that day—clouds on the horizon, the perennial suspension of dust from the desert, and some pollution from car fumes conspired to obscure the sun until it was already some way up in the sky. Nonetheless, the view was stunning, the haze making for a setting characteristic of this place. That the sun even appeared at all was a blessing we made sure to cherish. The missed spectacular dawn was a reminder of what Dubai does best—making the most of what we have, because man can only ever accept, and adjust to, what nature provides.

In Dubai, a metropolis rising right out of desert, there is an element of the desire summed up in the phrase made famous in the 1988 movie *Bull Durham*: 'If you build it, they will come'. But here, impressive skyscrapers are not just add-ons. They are part of a comprehensive national architectural ambition that is an essential, inherent feature of the city and the country. It is not just about one tower or one central business district. Here, stunning achievements of building design and construction reshape the whole look and feel of the city. The coastline is refashioned spectacularly like nowhere else, just as is the case with all the signature examples of infrastructure related to tourism and foreign investment, including highway networks, luxury hotels and office blocks.

Sand envelopes Dubai, creating a challenging environment for all except camels, four-wheel drive vehicles and helicopters. But what the city does with sand blows the mind, in a remarkable redefinition of what human design and engineering is capable of. The Burj Khalifa pierces through all preconceptions, a sensational spearhead at the vanguard of the UAE government's decision to diversify from an oil-based economy to one geared more to tourism and other services. Reaching for the top, indeed, is at the core of the UAE's brand essence.

In the race amongst nations to reach for the skies, the UAE is the forerunner, sprinting ahead by a long stretch. It has secured such a commanding lead that many other countries, to be sure, must have given up the chase, even those that at one time might have been within touching distance of similar architectural leadership. For any city, having an imposing and memorable skyscraper certainly adds something special to its skyline. If such a tower gains iconic status, it can even achieve immortality, with its image practically seared into the minds of most people on the planet. Such edifices need not be the tallest; they can simply be beautiful. One of the best-known, of course, is the Eiffel Tower of Paris, forever associating the French capital with immediately recognizable impressions, ideas and emotions of romance, culture and whimsical creativity.

Nonetheless, for buildings, height is the first obvious criterion. To build impressive structures is a human ambition that dates back to time immemorial. This is true of ancient peoples assembling the highest piles of rocks to honour the gods of nature, from large civilizations such as the Incas to the inhabitants of the tiniest Pacific islands and medieval societies across the globe, erecting the tallest steeples and domes for their places of worship.

Sand envelopes Dubai. But what the city does with sand blows the mind, in a remarkable redefinition of what human design and engineering are capable of. Reaching for the top, indeed, is at the core of the UAE's brand essence.

It is a yearning that is still very much alive. As at the first quarter of 2021, there are seventy-four buildings that reach a height of 350 metres (1,150 feet) or more, as assessed by their highest architectural feature. Of these, almost half—thirty-six—are in China, including Shanghai Tower, second-placed at 632 metres. Merdeka 118, a 118-storey, 644-metre skyscraper currently under construction in Kuala Lumpur, Malaysia, will become the world's second-tallest building when completed. This is one competition that seems to know no end, but at 828 metres (2,717 feet), the Burj Khalifa appears to have a comfortable lead for quite a while.

For the Burj Khalifa, all the effort that went into creating this architectural wonder is worth it, to say the least. At the start of our journey, on the Singapore Airlines flight from Changi Airport into Dubai, it was a delight to spot the Burj Khalifa amongst the sparkling lights of the city, although the dim light and the reflections in the plane window prevented any decent photo to be captured. The tower's unmistakable outline, which I had seen in photographs but not yet in real life, was striking. Later, on our last night in Dubai, we had an alfresco table at a Thai restaurant by the pond in the middle of Dubai Mall, with the Burj Khalifa right in front of us. The light show projected on to the whole facade of the skyscraper, the lights moving in sync with dancing jets from the fountain in the pond, was a snazzy accompaniment to our meal. Given what this skyscraper represents, as a marker of human

Something out of nothing: The Burj Khalifa represents man's capacity to conceptualize and create.

ambition, this sight can touch the spirit in a way that concrete structures seldom do.

The extremely high visibility of the Burj Khalifa has enabled it to be used also as a platform from which to project messages of global impact. This is done on festive occasions, and sometimes also to send signals of friendship with other nations, such as on these countries' national days, and also at other significant times. For example, images of China's flag were projected on to the surface of the tower in February 2020 to express solidarity with the government and people of China as the Asian nation was dealing with the worst of the outbreak of what was then called the 'novel coronavirus'. All in all, the Burj Khalifa represents man's capacity to make something out of nothing, and what an impressive symbol it is.

Imposing the ingenuity of man

If all nation brands were lined up to make a skyline, brand Dubai would be a skyscraper, for sure. We had visited Dubai once before—fifteen years ago, on a stopover on the way home from a trip with our friends Theng Hwee and Siew Yim to Syria, Jordan and Lebanon. That time in Dubai was a whirlwind drive-through, a four-hour city tour in the time between flights—just a tease of an appetizer to this fuller visit.

That stopover was when we only passed by the famous spice and gold souks, way past closing time, although the money changers were still open. We got a feel of the much-marketed coastline with its signature fancy hotels, including the rage at that time, the Burj Al Arab Jumeirah, billed then as 'the world's most luxurious hotel'. We knew then that to get a better feel of this city would require more time. Since Dubai—as one of the world's major air hubs—was a crucial and useful stop on a round-the-world itinerary, this would be our second chance to discover the city.

This time, we would also get to see items that we had missed previously, such as saffron, the scarlet seasoning and flavouring agent that is the world's most expensive spice by weight. With more time available, we would also get to step into places such as the Jumeirah Mosque, which is not only one of the few mosques in the UAE open to non-Muslims, but actually goes out of its way to welcome visitors. This is yet another aspect of the orientation of the UAE to forge connections with the rest of the world. This global outlook is part of the 'Open doors, open minds' programme of the Sheikh Mohammed Centre for Cultural Understanding. This centre was set up in 1998 because Sheikh Mohammed bin Rashid Al Maktoum, vice president and prime minister of the UAE, saw the need to reach out and educate expatriates and overseas visitors about the traditions and customs of the UAE.

As a traveller anywhere observes details of a destination, making comparisons with one's own home country is the most natural thing to do. For a Singaporean in Dubai, this becomes even more interesting and significant, because the one city that Dubai is often compared with is Singapore. Dubai can be said to be a more 'extreme' version of Singapore in some interesting ways. Both places share a similar broad trajectory of development from fishing village and desert to global city. The use of the word 'trajectory' is particularly apt, alluding to the arc, speed and sure path of the meteoric rise of both cities. The similarities between Dubai and Singapore are exemplified by the ease with which our Singaporean friends Rosebi and Netty, with their son Muhammad—whom we met for lunch at Wafi Mall—had settled into their new life in Dubai, just a few months after moving there for work.

One 'Singaporean quality' in Dubai is the vision, will and stamina to apply the ingenuity of man to overcome the challenges imposed by nature. Generally, no one can fail to be impressed by Dubai as an ultra-modern city that has emerged from what was just desert. Here too, there is the concerted political will and long-term vision, the daring to invest big in huge infrastructural projects and openness to welcome cutting-edge foreign investments on a global scale. Land reclamation in the sea is one of the most obvious outcomes, where Dubai has outdone Singapore or anywhere else by far, with projects such as the Palm Jumeirah luxury resort, part of a massive development that was planned to extend the shoreline by 520 km when completed.

Together with a few other places elsewhere around the world, the world map is still changing by the minute, as more sand is thrown into the sea to extend dry territory. For example, cities on China's coast reclaimed an average of 700 square kilometres of land—about the size of Singapore—every year from 2006 to 2010 for new houses, industrial

'Open doors, open minds': Jumeirah Mosque welcomes visitors as part of the UAE's programme to promote cultural understanding and forge connections with the rest of the world.

zones and ports. The scale in China might be more extensive, but Dubai still holds the edge in brand-building, for example, leveraging the pull of celebrities as brand ambassadors like the British footballer David Beckham, who owns property at Palm Jumeirah and opened the Adidas store in Dubai in 2015. The Emirates' association with global sport, especially football, through its airlines' sponsorships has won for itself much brand affinity for many years. For example, the English football club Arsenal not only has 'Emirates Fly Better' printed on every jersey but its stadium is also called the Emirates.

Just as with the central business district, when it comes to offshore real estate, Dubai, once again, reaches for the top. Nothing less than superlatives will do. Dubai's The World development (reclaimed islands shaped to look like the world map when seen from the air) and the Palm Islands are massive and distinctive enough to be the only man-made objects visible from outer space, other than the Great Wall of China.

When it comes to offshore real estate, Dubai, once again, reaches for the top. Nothing less than superlatives will do.

Since the global financial crash of 2008, however, the potential of these developments has once again become latent, an unfilled archipelago of white isles. The further extension developments of the Palm Islands have been held up, just like the 300 man-made islands of The World off the Dubai coastline. But the tides appear to be turning—cranes restarted work from 2014, although the much slower pace of development since then could reflect how these aspirations might have reached too far, at not quite the best of timings. And now, the pandemic has thrown another lorryload of spanners into the works. In the years ahead, to bounce back from the disruptions brought by Covid-19, in the most successful cities, aspiration's close sibling will have to be resilience. The pandemic is a reminder that the inevitable ups and downs in the economy, society and politics require much patience and resourcefulness to overcome.

Another aspect of similarity between Dubai and Singapore is the reliance on imported augmentation of the labour force at all levels. In Dubai's population of more than nine million, nearly nine in ten people are expatriates—one of the highest ratios anywhere in the world. On the street, the difference in diversity is obvious. Everyday Singaporeans are available everywhere to give visitors directions to, say, the nearest hawker centre or the furthest restaurant. But UAE nationals are more difficult to interact with, as there usually appear to be so few of them around in

Three faces of Dubai: Anti-clockwise from top—Locals traversing Dubai Creek and welcoming passing foreign visitors on motorized water taxis called abras, expatriates enjoying the highest standards of cosmopolitan creature comforts, and migrant communities, of whom South Asians form the largest group at 60 per cent of the population.

the tourist areas. Of course, many other cities contain within themselves a multitude of ethnicities, nationalities and backgrounds. But the UAE takes this to a whole other level. The 10-to-90 per cent proportion of locals to foreigners is a statistic cited by our guide Abdul, himself an Egyptian. In Dubai, some communities of foreigners in sectors such as construction live in a kind of 'parallel universe', quite apart from the society in general.

Many tour guides in Dubai are from Egypt, a traditional source of labour for this sector, valued for their familiarity with Arab culture, fluency in English, comfort with guests from English-speaking countries, and being well-trained in Egypt's tourism industry. As these guides give their explanatory introductions to the sights along the way, their commentary establishes a certain rapport with their audiences, both guides and tourists being fellow outsiders setting out to appreciate this city and country better.

With Dubai's oil reserves now running low—down to a single digit of contribution to national revenue—other industry sectors such as tourism are gaining heft. But, as always, the prospects for any sector, or job, are a function of demand and supply. Older UAE nationals would know this dynamic better than those who are younger, if they recall the rise and fall of Dubai's pearling industry in the more distant past.

Western commentators sometimes try to put a spotlight on negative aspects of the working conditions of foreigners in the UAE, but for a place like Dubai, this is one model that works to meet the labour needs in sectors such as construction, to enable the realization of the city's economic aspirations. And the people who are here look happy enough, able to make a more-than-decent living, even if they are ensconced in their own primary social circles most of the time. Dubai has a national Happiness Agenda to make the city the world's happiest. It looks as if there is, at least, a foundation of general contentment to build on. A deeper search for what was at stake in this quest for happiness led us on to Abu Dhabi.

Abu Dhabi: Connecting everything in the universe

Just over an hour away by road from Dubai, Abu Dhabi, the UAE's capital and second-most populous city after Dubai, reaches for heights of human aspiration in different ways. 'Everything in the universe is within you'—this literary quotation adorns one of the glass walls near the main entrance of the Louvre Abu Dhabi museum, one of two sights that we visited on our day trip to Abu Dhabi.

To most foreigners, this might seem like a line plucked from some obscure poet with a Middle Eastern-sounding name. Some would be surprised to find out that its author, Rumi, is actually quite famous in the West, his poetry having inspired many international celebrities, including the rock band Coldplay's frontman Chris Martin, who has incorporated Rumi's verse into his music. Aptly, this line of verse also encapsulates the UAE's aspirations to expand the Muslim worldview and to forge connections with the rest of the globe.

Jalāl ad-Dīn Muhammad Rūmī (1207–73) was a thirteenth-century Persian poet, Islamic scholar, theologian and faqih (expert in jurisprudence), and Sufi mystic, originally from Greater Khorasan (today

Everything is within you: A line from the thirteenth-century Persian poet Rumi greets visitors soon after entering the Louvre Abu Dhabi, reflecting a worldview that is at once Islamic and globalist.

part of Afghanistan). More popularly, he is known simply as Rumi, and his literary and inspirational influence has transcended national borders and ethnic divisions for the past seven centuries.

Coldplay's Chris Martin revealed in an interview that, in 2015, when he was going through a divorce from the actress Gwyneth Paltrow and feeling down, a friend gave him a book to lift his spirits, a collection of poetry by Rumi, translated by Coleman Barks. It is true that such Western translations of Rumi's poetry have downplayed its Islamic elements so much as to be accused of almost total 'erasure'. But anyone who is drawn in to read up more would discover the poetry's deeper, religious inspirations. At least, to begin with, there is in Rumi's verse that initial cross-cultural window, or bridge, to invite foreign audiences into this world.

The poem 'The Guest House'—which inspired the song 'A Head Full Of Dreams' from the 2016 Coldplay album—is from Rumi's *Masnavi*, a six-book epic poem of 50,000 lines that he wrote towards the end of his life. Rumi is often described as a bestselling poet in the US, with millions of his books sold in translation. He has also been a source of help on the spiritual journeys of other celebrities, mostly in the music industry, including Madonna and Tilda Swinton, some of whom have also incorporated his work into theirs. The singers Beyoncé and Jay-Z named one of their daughters Rumi.

At the heart of the UAE's positioning and promotion of a global openness, which is its nation brand essence, is a reaching down to the core of what binds all human beings.

The line 'everything in the universe is within you' speaks of how everything in the world is never separate but always connected. Every ego, or natural or spiritual state, is a part of the divine, and there is a part of the divine in every person. What accounts for Rumi's popularity is the unifying power of deep humanity in his words. It is this essential thread of connection across geographies and ideologies that is at the heart of the UAE's positioning and promotion of a global openness, which is its nation brand essence. It is a reaching down to the core of what binds all human beings, far below what divides at the surface, whether it be factions of faith or camps of colour.

The Louvre Abu Dhabi itself is a shrine to shared humanity. It is an example of how to fulfil the brand advisor's dream of living up to the museum's official brand tagline: 'See humanity in a new light'. On its website, a self-imposed question asks: 'What does it mean to be universal?'. The answer, for this museum, is that 'it means focusing on what unites us'.

The twain shall meet: At the Louvre Abu Dhabi, women from near and far consider artistic works from Europe and the rest of the world.

The museum has a billion-dollar agreement with the French government to use the Louvre name and to exhibit artworks from twelve French cultural institutes, including the Musée du Louvre, the Centre Pompidou and the Musée et Domaine National de Versailles. The initial temptation could have been to stock the rooms with artefacts of French culture, but instead, while the selected items are drawn mostly from collections held in France, they make an excellent effort to represent the whole world.

Instead of the conventional way of presenting pieces by geography, time period or technique, the aim here is to portray the strands of commonality across civilizations. For example, carved figurines of horsemen from the quite separate civilizations of Iran, Cyprus and China are placed side by side in the same glass case, inviting the viewer to consider the similarities amongst these ancient cultures. Such juxtaposition jolts all manner of preconceptions, nudging the viewer towards a deeper reassessment of the sense of the world that they grew up with. For example, in one display, three statuettes are placed in a row to highlight the similarity in their posture of holding their arms together as if in prayer. They are from Greece (2700–2300 BCE), Syria (2500–2400 BCE) and Gabon (1800–1900), and together, they depict the universal human desire to connect with the celestial world.

This mode of curation is what positions the Louvre Abu Dhabi as what has been called the first 'universal museum' of the Arab world, one that tries to explain the whole world, not just particular regions, through art history. For the local visitors, what this museum brings is a significant value-add to perceptions of the rest of the globe, opening up vistas especially for an audience more used to the customs and conventions in Muslim art of avoiding depictions of human or animal figures.

The galleries culminate in a striking work in the last room—*Fountain Of Light* by Ai Weiwei, the artist and activist from China who moved in self-exile to Berlin, Germany, and then to Cambridge, England. Like a giant chandelier inverted onto the floor, its shape recalls the Tower of Babel in the sixteenth-century paintings of the Dutch artist Pieter Bruegel the Elder. Ai's work is actually based on a Russian work from 1919, Vladimir Tatlin's *Monument To The Third*

Upturned vision: Is the artwork *Fountain Of Light* by Chinese artist Ai Weiwei a comment on a fragile utopian dream?

International, a twisting building for the proposed headquarters of the Comintern, the international organization that aimed to bring about world communism, which was never built. An aborted celebration of a failed utopia, the Russian project can be read as a comment on a communist regime ironically infatuated with opulence, seeking a tribute that is ultimately fragile and unsustainable. Ai's work, referencing the comparative context of China, raises similar questions about utopian experiments and failures.

For some viewers, Ai's work is enjoyed more for its Instagram-worthy imagery and the superficial association with fame that it affords selfie-takers. Some of these onlookers might no doubt be oblivious to the deeper significance of such artworks, conveyed in the often-implicit social and political commentary of art. But actually, having Ai's work displayed at all, not to mention placed in a position of privilege as the 'finale' piece of the museum, reflects the admirable openness of the Louvre Abu Dhabi, and of the UAE too, to self-questioning—the chandelier's connection with light shines greater significance on the link to the museum's own slogan: 'See humanity in a new light'.

Game-changing scale and a globalist worldview are also hallmarks of the other major sight in Abu Dhabi—the Sheikh Zayed Grand Mosque.

Game-changing scale and a globalist worldview are also hallmarks of the other major sight in Abu Dhabi—the Sheikh Zayed Grand Mosque. This mosque can host more than 40,000 worshippers and has three Guinness World Records: for the world's largest chandelier, dome and handwoven carpet of these types. The project was launched by the late UAE president Sheikh Zayed bin Sultan Al Nahyan, who wanted to establish a structure that would unite the cultural diversity of the Islamic world with the historical and modern values of architecture and art. His final resting place is located on the grounds adjacent to the complex.

The UAE's international outlook is reflected also in the floral designs that adorn the floors, walls and pillars of the mosque. Made from more than thirty million pieces of Italian marble, they were created by British artist and illustrator Kevin Dean. And the market test for the mosque seems to have been passed with flying colours: The mosque was ranked by travellers as the world's third favourite landmark in 2019, according to the travel website TripAdvisor, and second for two years running before that, in 2017 and 2018, out of more than 700 landmarks in over eighty countries. This is amazing, given the number of major tourist landmarks around the world.

As part of the approach to open up the Muslim world to people from other cultures, the Mosque loans traditional abayas (robes) to visitors, especially women, to align with Muslim norms of modesty. After my wife put on her loaned abaya, from behind, she looked like almost every other female visitor. Every time we met up after our separate explorations of the mosque's vast interior, I had to walk ahead

Welcoming outsiders: Abu Dhabi's Sheikh Zayed Grand Mosque loans traditional abayas (robes) to visitors, in the spirit of sharing aspects of Muslim culture.

to see her face before speaking to her, because I could not, for example, first touch her arm to get her attention.

The UAE's relationship with the outside world is summed up in the quotation from Rumi at the beginning of this chapter—'Everything in the universe is within you'—which is actually followed in the poem itself by the words: 'Ask all from yourself'. Indeed, one way of looking at the UAE's international aspirations is that whatever the planners, brand advisors and marketers might offer, ultimately, everything hinges on resonance, reception and reciprocity. Those in charge of place branding may present a certain positioning, but if this does not resonate with the people and is not well received, then everything will ultimately be in vain if audiences do not reciprocate with their own personal appreciation and self-reflection to 'ask all from themselves'.

Dubai ranks as the world's seventh most-visited city, according to the Euromonitor International's data on 600 cities around the world, so there must be some resonance there. The reception from visitors has also been mostly positive, judging by the social media popularity of 'bucket-list icons' such as the Burj Khalifa and the Grand Mosque, which many people hope to visit at least once in their lives.

Reciprocity is where what is asked of oneself is for a spirit of receptivity from all sides. Just as when UAE nationals and visitors from other parts of the Middle East encounter artworks from across the world, their Muslim worldview is opened up to other possibilities, so too, visitors are invited to maintain a similar generosity of openness in the other direction. Everyone is invited to re-examine the interface between the spiritual and the secular, and in the process of that search, to keep the door open to embracing something new and more unifying. This, in the end, is the greatest value in the best elements of spirit that the UAE represents and projects.

Expo 2020 and the future

With Expo 2020, a World Expo that calls everyone to visit to 'see the best of the world in one place', Dubai sustains its key brand attribute of always reaching for the best-in-class in its initiatives of global outreach. The Expo, featuring almost 200 countries presenting their nation brands in Dubai, was originally scheduled for six months from October 2020, but was postponed for a year due to the Covid-19 pandemic. The show has three main themes: opportunity, mobility and sustainability. These three themes address the top three global challenges of the present times: social division, technological disruption and environmental degradation. The Expo's hashtag keyword, '#hayyakum', an Emirati word meaning 'welcome', is well aligned with the UAE's nation brand positioning.

What gives the themes added substance in relation to the host location of this expo is that the UAE has been walking the talk in all these three areas. It has been very active online and in deploying technology to forge social cohesion and encourage its own people to look outwards, and to connect with other cultures. It has also invested a lot to enhance its capacity to manage the environment while reshaping it, including exploring new energy sources such as a solar project that is slated to be, of course, the world's largest.

The UAE's noted contributions to the world may, in time to come, extend to include the promotion of happiness. Dubai has an initiative called the Dubai Happiness Agenda, which has sixteen programmes under four themes (to meet the people's affective, basic, cognitive and deeper needs). These themes sum up eighty-two projects to be set up in the city, with an aim to make the city the world's happiest. This initiative takes up the lead originally set by Bhutan with its Gross National Happiness (GNH) model, which goes beyond focusing on gross domestic product to include emphasizing citizens' welfare in addition to economic growth. This more holistic way of long-term planning in public policy has also influenced other countries such as New Zealand, which announced its much-praised 'well-being budget' in 2019, designed to go beyond boosting the economy to also addressing the growing disparity between the haves and have-nots.

Bhutan's GNH framework has been used by its government since 2008 to analyse any proposed policy to ensure that it best serves the collective well-being of the population. This is done by considering factors under four 'pillars': sustainable and equitable socio-economic development, environmental conservation, preservation and promotion of culture, and good governance. This Bhutanese concept became the impetus for the United Nations' promotion since 2011 of happiness as a

The way ahead: 'Shaping the future' is the slogan of this poster for the Dubai International Book Fair 2018, an outlook taken to new levels with Expo 2020.

fundamental human goal for all nations. A more holistic framework was adopted, in contrast to the earlier, more narrowly focused Sustainable Development Goals. This would then enable the world's governments and peoples to work better together to protect the environment as part of enhancing the entire gamut of aspects of quality of life for human societies everywhere.

Advancing a similar approach, the UAE has exercised international leadership in advocating happiness, convening interested governments to meet annually in Dubai at the World Government Summit and to join the Global Coalition for Happiness and Wellbeing, formed in 2018 in the UAE with five other founding countries: Costa Rica, Mexico, Portugal, Kazakhstan and Slovenia.

Happiness, in my opinion, should be at the top of the agenda for any government anywhere. The shock of Covid-19 has made this idea even more imperative. Weighing up every initiative through a comprehensive lens of multifaceted considerations is the only way to predict, pre-empt and preclude the unwanted negative consequences of any policy, and thus serve the best interests of the people. It is no surprise that the UAE, with its openness to the best ideas available, should have adopted the prioritization of happiness too as part of its vision for the future. Whatever one might say, the UAE is one country that keeps reaching for the top—figuratively and literally.

Oman

Slivers of sanctuary

In the haiku poem and ink sketch in this haiga artwork, the turtles that keep returning to nest on the beaches of Oman act as an extended metaphor for the country itself—how it never fails to find its way back to shore every time. What maintains its sense of direction and focus is the nation's heritage and traditions, the essence of this home of an ancient seafaring people, seeking to swim with the currents of modernity to explore and harvest the bounty of the oceans.

OMAN

Against the tide

going with the flow
is surer, with a homing
body compass stowed

許木松
2020

Sea of serenity: A turtle sanctuary on Barr Al-Jissah Bay in Muscat is ensconced off the Gulf of Oman and the Arabian Sea—a blend of effort, with some care, to nurture tradition while negotiating its interfaces with modernity.

Safe in Bedouin hospitality

'My name is Saif,' said our guide, as he met us at Muscat International Airport in the capital of Oman, on arrival from Dubai. The homonym with the English word 'safe' was immediately obvious—especially in the context of travel, applying more to crime than to catching any virus in the pre-Covid-19 world. He too picked up on the pun at once, probably not for the first time. 'I will do my best so you will have a safe and enjoyable trip,' he assured us, with a smile. And indeed, that was how we felt throughout our time there. In Oman, we always felt safe, with Saif.

Saif was, in many ways, an embodiment of the modern Omani—living close to ancient heritage, yet open and welcoming to what modernity might bring in from the rest of today's world. This harmony of contrasts is captured in dressing, not just in the items of clothing themselves, but in the bearing and confidence with which such apparel is carried. This attribute was there in the dishdasha that he wore—the Omani national dress for men, an ankle-length, collarless, usually white, gown with long sleeves, complete with a cap with multicoloured threading, called a kummah. The Omani style was there also in his sunglasses—the only concession to modern, Western 'fashion', because it is essential to shield his eyes from the incessant, ubiquitous glare of the sun. Never did I feel that he wore the shades also to 'look cool', unlike, say, air force pilots who might wear shades even indoors. His demeanour was aligned with his Omani worldview: interested, and invested, in connecting with the modern universe, yet always blended with an essential character rooted in 'Bedouin values'.

The Bedouins are a grouping of nomadic Arab people who have historically inhabited the desert regions in North Africa, the Arabian Peninsula, Iraq and western Asia. The English word 'bedouin' is said to have come from the Arabic 'badawī', which means 'desert dweller', and is traditionally contrasted with ḥāḍir, the term for sedentary people. And even though the Omani people of today are more ḥāḍir in lifestyle amidst a greater wealth of creature comforts, there is still a fascinating combination of both facets in Oman.

This is true, particularly, of Omanis who have a lot of contact with the outside world, such as those in the tourism industry. In fact,

Tradition lightly worn: For Omani men, a dishdasha with a cap called a kummah, seen here in the late morning sun, against a backdrop of the rocky landscape surrounding Nizwa town.

because they spend so much time sharing and explaining knowledge about their country, they may be all the more rooted, more ḥāḍir, even as they remain badawī in nomadic spirit, in their sense of adventure and exploration.

> The motivation in Oman is always to help others to understand Muslim culture, to bring others in, while working just as hard to encourage their own people to expand their mentality and worldview, so as to build more bridges with the rest of humanity.

Some say the key Bedouin values are honour, pride and generosity. Saif demonstrated all three. His very manner and bearing exuded honour, for want of a more accurate term. We remarked on how neat his appearance always was, in his starched dishdasha, even amidst the sweltering heat on a walking tour. We never saw a drop of perspiration on him, with the low humidity of Oman no doubt a factor. In any case, he revealed to us that he always carried in his car a spare outfit in case some 'wardrobe mishap' should happen to the one he was wearing, like a stain or a tear. He spoke with pride of how Oman, as a country and people, is focused on aligning with modernity wherever it can fit in with the foundations of heritage and faith. To achieve this, generosity—a key ingredient—was there in his own demonstrated spirit of openness to new, foreign ideas.

These Omani brand attributes, embodied by Saif, bear some similarity to those of the UAE. There is the obvious common thread of a Muslim nation seeking to adopt and sustain a welcoming stance towards the rest of the world. There are many positive qualities of wanting to accept, even to embrace, diversity. The motivation is always to help others to understand Muslim culture, to bring others in, while working just as hard to encourage their own people to expand their mentality and worldview, so as to build more bridges with the rest of humanity.

The cosmopolitan openness to the world that is a feature of modern Oman was not always the dominant posture towards outsiders. In ancient times, survival was a matter of constant attention to self-defence.

Living heritage: Centuries-old lifestyles and livelihoods are sustained by these daytime residents of the ancient fort and souq (market) of Nizwa town.

For example, in Nizwa town, which we visited on a day trip from Muscat, structures such as the old Nizwa Fort bear testimony to the longstanding need to ensure security. Inside these fortifications, Nizwa is a living museum, preserving a substantial portion of Omani culture and heritage. The aqueducts that run around the perimeter of the fort and souq (market)—an ancient engineering system that also channelled water to the surrounding farming villages—are still bearers of a vital lifeline of sustainability. The souq still houses centuries-old practices of local agricultural trade and their associated commerce, symbolized by the traditional snack of mutton skewers that we were treated to at the gates of the souq before we entered, a nourishment from nature prepared by man for sustenance.

Once a seafaring power in centuries past, Oman is the Arabian Peninsula's understated cultural gem. Akin to a global port, like Singapore in some ways, Oman's key position on the ancient trade routes allowed it to absorb influences from India, Persia and Zanzibar as well as Portugal and Britain. As the oldest independent state in the Arab world, its history of self-reliance might go some way towards accounting for the sense of self-assuredness that one can discern across all aspects of its culture, as well as being projected by people like Mohammed, my friend from graduate school in the US, who is the epitome of a gentleman and is comfortable interacting with people from any other culture or background. Unfortunately, he was out of town during our visit, so our mini-reunion in Muscat happened only in spirit. But it is only in spirit, in any case, that human connections are forged across cultural and other barriers.

Bridging differences is also the key to Oman's traditionally non-interventionist stance in foreign policy, playing a diplomatic role amongst the Gulf States, even as armed conflict surrounds it in neighbouring places such as Yemen, buffeted by the influence of major powers such as Saudi Arabia. It is a very well-used cliché, but Oman truly is like an oasis of calm and peace in a region where one can easily be lulled by the soporific desert heat, when in fact, at any moment, the winds can stir the sand to rise into a gathering storm.

Within the country, staying neutral has also helped ensure stability. Oman's self-sufficiency also appears to have been a bulwark against cultural infiltration, unlike other places in the region, where incursions from external sources have reached a bit further. For example, in Morocco, fake Ray-Ban sunglasses made in China enable locals and some tourists to put on Tom Cruise's cool look from the movie *Top Gun* at just a fraction of the normal price. But, unfortunately, this accessory can start to fall apart after a week of wearing. By contrast, Oman's heritage, perhaps less infiltrated by international materialism, can hold much more firmly.

Cocoons for nature: The turtles of Oman

A wondrous sliver of sanctuary for nature in Oman are the turtle nests on its beautiful beaches. There are just five nesting sites in the country, including beaches with public access such as Ras Al Jinz Turtle Reserve, and private beaches such as at Barr Al Jissah, where hotel guests do not need to hike anywhere to see the nests. This is like a lazy sunseeker's

Turtle season: These determined amphibians paddle on to Oman's beaches to nest, in an endless struggle to perpetuate an endangered species.

dream form of ecotourism. The turtle nests are marked out on the beach, a beachball's throw away from the hotel deckchairs on the sand. To prevent anyone from trampling on the buried eggs, wooden stakes are set up around each spot bearing information about the nests there. These roped-off nests are scattered around the beach, interspersed safely in between the beach deckchairs, and not even interrupting any sea sports activities.

Every year, approximately a hundred turtles return to the resort's beach to lay their eggs, with each nest containing up to 120 eggs. The turtles swim there from the Arabian Gulf and as far away as the Red Sea. Two species visit at this particular beach: the endangered green turtle and the most frequent visitor, the critically endangered hawksbill turtle. The other three species that visit Oman are the loggerhead turtle, the Olive Ridley and the leatherback.

The period from January to August is the main turtle season. We visited in April, about a month too early to see turtle eggs hatch, but if you are lucky to be there at the right time, it would surely be an awesome sight. In 2018, about 3,126 turtle hatchlings were produced from ninety-five nests laid out along the beaches of Al Husn and Barr Al Jissah Resort and Spa, from fifty-one hawksbill turtle nests and forty-four green turtle nests.

The resort has its own 'turtle ranger', Mohammad Al Hassani, who leads a team to look after the turtles, ensuring that the nests are well-monitored and protected, and the turtles and their eggs are shielded from harm. This means keeping at bay potential predators including foxes, crabs and birds.

In more ways than one, Oman itself is like one big metaphorical beach with turtle nests—a sanctuary of nature, nestled amidst fragile beauty, that depends on its interaction with the ocean of the outside world, which is potentially treacherous but undoubtedly necessary to survive and thrive.

A global stage: Opera in Muscat

Just as turtle eggs store centuries of DNA within their thin shells, Omani culture is like the DNA, the life force, of Oman. Sultan Qaboos bin

Said Al Said had long declared it his goal to create a peaceful, inclusive modern state without neglecting its 'glorious heritage' and 'precious evidence of its past'. For such inclusivity, there is no more conducive arena than culture and the arts, and the Royal Opera House in Muscat, built on the royal orders of the Sultan, is a wonderful place. Here is the home of the only orchestra in the Middle East composed entirely of musicians from the region. The performing space has hosted performers from around the world, from the renowned cellist Yo Yo Ma to the London Philharmonic Orchestra. The only other opera house in the Arab world is in Cairo, Egypt.

The Royal Opera House is a distinctive example of Arab architecture, with exterior arched windows and an ornate entranceway. By international standards, this opera house is modest in size, about one-third the size of the Kennedy Center in Washington, DC, in the US. But its significance and potential are massive in this part of the world. Built in marble and wood, the interior of the opera house is a lavish setting, with wood decorated in gold and handmade ornaments. But while the setting is ancient, the facilities are very modern. Each red seat has a touch screen so that patrons can choose a language in which to see surtitles. Enriched with such culture, audiences' mindsets can only become more modern. In this way, the Royal Opera House is a beacon, like a lighthouse on Muscat's beaches, just as Oman itself is an oasis of welcoming openness on this peninsula.

Just as turtle eggs store centuries of DNA within their thin shells, Omani culture is like the DNA, the life force, of Oman.

Oman's intended nation brand positioning, respecting the perch of tradition and reaching out over the cusp of modernity, was, to me, perfectly symbolized by the opera that was the opening premiere production for the Royal Opera House in 2011—*Turandot*, an opera in three acts by Italian composer Giacomo Puccini. Director Franco Zeffirelli's production in Muscat was conducted by Spanish tenor Plácido Domingo.

The prominent exhibits on permanent display in the lobby of the opera house are a reminder of the cultural statement of this premiere

World stage: *Turandot,* the Royal Opera House's opening show in 2011, is a symbol of Oman's orientation to globalization.

production, as well as of the opera house, and, indeed, of Oman's cultural and globalist ambitions. Having this display at the opera house was an excellent idea. Walking around the lobby and viewing the exhibits, we were almost transported back to the evening of the premiere.

The original tale of *Turandot*—of a princess who has sworn that no man shall marry her unless he can answer three riddles—is actually linked to Arabia through its Persian origin story. It is based on one of the seven stories in the epic poem *Haft Peykar*, a work of the twelfth-century Persian poet Nizami Ganjavi. The poetic work, completed in 1197, is also known as *Bahramnameh (The Book of Bahram)*. It tells the story of Prince Bahram Gur, ruler of Sassania, the last kingdom of the Persian Empire (224 CE–651 CE) before the rise of Islam, and his progress towards wisdom. The tale is enlivened by many adventures and by the seven love stories told to the prince by his brides. The Persian title *Haft Peykar* can be translated literally as 'seven portraits' with the figurative meaning of 'seven beauties'.

Turandot sets the stage for Oman's new relationship with Asia, and by extension, with the rest of the world.

Nizami aligned the seven stories with the seven days of the week, the seven colours of the rainbow and seven corresponding planets. The story of Turandot is for Tuesday, being told to Prince Bahram by his companion of the red dome, associated with the planet Mars. In the very first line of this tale, the protagonist is identified as a Russian princess. The title of the opera is based on Turan-Dokht (daughter of Turan), a common name used in Persian poetry for Central Asian princesses.

The opera's narrative is set in China and involves Prince Calaf, who falls in love with the cold Princess Turandot. To obtain permission to marry her, a suitor has to solve three riddles; any wrong answer results in death. Calaf passes the test, but Turandot still refuses to marry him. He offers her a way out: if she is able to learn his name before dawn the next day, then at daybreak he will die.

The opera is what some might call an 'Oriental fantasy', complete with Caucasian actors dressed in the costumes and make-up of ancient China. For many years, the government of China forbade performances of *Turandot* because it said the opera portrayed China and the Chinese unfavourably. In the late 1990s, the Chinese government relented, and in September 1998 the opera was performed for eight nights as *Turandot At The Forbidden City*, complete with opulent sets and soldiers from the People's Liberation Army as extras. It was an international collaboration, with Chinese director Zhang Yimou as choreographer and Zubin Mehta as conductor.

As with the opera *Madama Butterfly*, set in Japan, Puccini strove in *Turandot* for a semblance of Asian authenticity (at least to Western ears) by using music from the Asian region. Up to eight of the themes used in *Turandot* appear to be based on traditional Chinese music and anthems, and the melody of a Chinese song titled 'Mò Li Hūa (茉莉花)', or 'Jasmine', is included as a motif for the princess.

These connections with China bring this aspect of the Royal Opera House's origins and cultural meaning into contemporary relevance. In the first half of 2019, the number of Chinese tourists arriving in Oman increased to around 64,000, about 50 per cent more than the previous year, as the country joins practically every other actively ambitious tourist destination in seeking to take a slice of the pie represented by the estimated 150 million Chinese tourists travelling abroad each year in pre-Covid-19 times. In this way, Oman—with its longstanding political stability, relative social openness and orientation to globalization— would be a 'welcoming sanctuary' for travellers to this Middle Eastern region. Fittingly then, *Turandot* sets the stage for Oman's new relationship with Asia, and by extension, with the rest of the world.

Self-determination and social cohesion

Conception of divinity defines everything else. Oman is one of only two sultanates—the other being Brunei Darussalam—so naturally, faith plays a dominant role in public life, as well as in the relationship between government authority and ordinary folks. Oman's national characteristic and brand attribute of self-determination has its roots partially in the fact that the Omani people practise their own version of Islam—the Ibadi school, which is separate from the more famous schools, Shia and Sunni.

This branch of Islam is also found in parts of Algeria, Tunisia, Libya and East Africa. In Oman, amongst the points of difference with the majority of the Muslim world is that, in the Ibadi school, there is the belief that it is unnecessary to have one leader for the entire Muslim

All is calm: The Sultan Qaboos Grand Mosque takes the general quietude of Oman up to another level.

world, and if no single leader is fit for the job, then Muslim communities can rule themselves.

This is unlike both the Sunni belief in a caliphate (government under a caliph, an Islamic spiritual leader who claims succession from Prophet Muhammad) and the Shia belief in the Imamah, the doctrine that imams (worship or community leaders) are the central figures of the ummah (Islamic community). In Oman, such religious interpretations are informed by self-restraint in practice and openness to foreign influences. These values are officially espoused, articulated and promoted. For example, one message that we saw on a large banner greets anyone entering the airport. Signed off by the Sultan, it reads: 'Oman, the civilisation: Was and still living within the approach of tolerance, moderation and civilised coexistence.'

Such societal cohesion is fostered also by other aspects of society, including the role of grand architecture in embodying, and as a focus of, political power. The primary monument of this in Muscat is the Sultan Qaboos Grand Mosque. Commissioned in 1992 by Sultan Qaboos bin Said al Said, construction took almost seven years, in time to celebrate thirty years of his reign in 2001. Built from 300,000 tonnes of Indian sandstone, the mosque has a main musalla (prayer hall) that can hold more than 6,500 worshippers, while the women's musalla can accommodate 750 worshippers. The outdoor areas have additional space—outer paved grounds, an interior courtyard and passageways—making a total capacity of up to 20,000 worshippers.

One could say that mosques are the equivalent of skyscrapers in the religious domain, and there appears to be a similar quest for scale, to build a place of worship that would be the proportionate equivalent of Dubai's Burj Khalifa tower, if one had the means. The main design feature of the Grand Mosque's interior is the prayer carpet that covers the floor of the main prayer hall. It contains 1.7 million knots, weighs 21 tonnes and took four years to produce, a modern cultural artefact of the region that brings together the classical Persian Tabriz, Kashan and Isfahan design traditions.

The details are all stunning. The main carpet used to be the largest single-piece carpet in the world but was overtaken by the carpet that adorns the Sheikh Zayed Mosque in Abu Dhabi, in the UAE. In the same way, the chandelier above the prayer hall—14 metres tall, with 600,000 crystals and 1,122 halogen bulbs, and made by an Italian company—was the world's largest chandelier until it was also displaced by the same Sheikh Zayed Mosque.

Above all: The chandelier above the prayer hall of the Sultan Qaboos Grand Mosque in Muscat was once the world's largest.

Some visitors—inevitably comparing the two mosques of Muscat and Abu Dhabi—suggest that the one in Oman is 'more authentic'. Aesthetic considerations are one main arena of 'competition' between the two structures. The beauty of nature is seen to be celebrated more artistically in the Muscat mosque, such as in the palette of twenty-eight colours in varying shades used in the décor in the Muscat mosque, mostly derived from traditional vegetable dyes. People who see this mosque as superior might be projecting their own preconceptions and tastes onto observed reality. Whether this assessment is true might, in itself, offer another line of spiritual enquiry—the possibility that authenticity, like spirituality, is always in the eye of the beholder.

The one-upmanship behind such feats of construction is such that, once a structure is overtaken by another somewhere else, the game changes, whether one likes it or not. And then, it is only when the dust of the race subsides, and the last fallen fragment of figurative confetti is swept up from the celebratory launch of the new edifice, that one can start to settle down to a more essential purpose for the building. And then the focus turns to where it should have been to begin with—to seek the serenity that such places of worship are for, in any case.

Sentinels of a Portuguese history

The heritage architecture of Oman highlights an intriguing aspect of ancient times—how the Western concepts of castles and fortifications of brick-and-mortar made their way from Europe and still stand today as silent, often empty, sentinels of a history long past. Examples include the Roman relics across places as far-flung as Syria and Lebanon. Many of these architectural features, as well as the concept and practice of fortified defence, complete with moats and drawbridges, can also be found in Asia. In the castles of Japan, the architecture shows a distinctly Japanese character, not least in terms of roof detail, gardens and interior decor.

Oman has the good fortune to share a slice of this heritage. Muttrah Fort, in Muscat, shares a history with the other great Portuguese coastal strongholds of Bahrain, Diu and Goa in India, and Mombasa in Kenya, being directly linked to the first Portuguese Empire, which was born during the Age of Discovery, the period from the early fifteenth century into the early seventeenth century. This was when European ships traversed the world to uncover fresh trading routes and sources of treasures and raw materials. Muscat had been in Portuguese hands since 1507, when a fleet under the Portuguese general Afonso de Albuquerque attacked the city, destroyed it, and then returned soon after to occupy it.

The maritime expertise of Portugal's seafarers of yesteryear enabled them to round the Cape of Good Hope (in what is modern-day South Africa) and find another way to access the lucrative spice trade of Asia, which until then had been monopolized in Europe by the Venetians (who worked exclusively with their trade counterparts in Egypt). The Portuguese were quickly able to establish themselves throughout the coastal areas of the Arabian Peninsula, including Oman, with parts of the Sultanate occupied by them for 145 years. It was in this era that Muttrah Fort was constructed.

Anchor of memories: The promenade of Muttrah in Muscat recalls a great seafaring past, seen at its most captivating at sunset.

The Muttrah Promenade is stunning at sunset, an ambience we enjoyed on a very pleasant stroll one evening. It was a superb way to soak in the atmosphere of this ancient waterfront, framed by golden hues in the sky, after an early dinner at a waterfront-facing restaurant of shuwa, a classic Omani celebration dish of marinated meat on skewers roasted in an underground firepit for more than six hours. Relishing this repast, we were reminded of the skewers we had enjoyed outside the Nizwa souq, a humbler snack for common folk, but almost as satisfying.

The whole promenade is rich with history, intricate as the patterns of Omani mosaic tiles. Widely recognized as icons of Muscat, the forts of Jalali and Mirani are situated in Old Muscat and—despite not being open to the public, except for some external structures—continue to fascinate visitors from around the world. Originally built by the Portuguese as part of Muscat's coastal defensive system against attacks by cannon fire, their usage continued to evolve with time, after advancements in the technology of war made the original architectural purposes redundant. Later, these two forts served as jails and garrisons, and also as royal residences and houses.

On the ramparts of history: The ancient Portuguese fortification of Al Jalali Fort in Old Muscat.

Jalali and Mirani are nearly always mentioned together, given their close proximity to one another and their rhyming names. No one knows for sure who christened them: one theory is that Jalali refers to the Arabic word jalal meaning 'elevated', while Mirani is a corruption of the Portuguese word mirante, meaning 'lookout', which also makes sense. Local legend, on the other hand, credits their names to Miran Khan and Jalal Khan, two soldiers said to have fought to capture these objectives during the last battle between the Portuguese and Omanis for Muscat in the sixteenth century.

If it is true that construction of Mirani Fort began in 1507, with Jalali built twenty years later, it shows the long duration of forts as key elements of maintaining power in Omani history. For example, Nizwa Fort was built in the 1650s during the Yaruba dynasty, which ruled Oman between 1624 and 1742, after expelling the Portuguese from coastal strongholds in Muscat and united the country. Nizwa Fort, in the Omani town of Nizwa, one and a half hour's drive from Muscat, is one of Oman's most visited national monuments. It was the administrative seat of authority for the presiding imams and walis (governors) in times of peace and conflict. The main bulk of the fort took about twelve years to complete and was built above an underground stream.

Looking over the forts' ramparts and through the lookout 'windows', it is clear that Oman's forts are powerful reminders of the significance of the towns they are located in, surviving through turbulent periods in Oman's long history. They served as strongholds against raiding forces that craved natural wealth and strategic locations at the crossroads of vital sea routes. The dizzying heights of many Omani forts and castles and the complexity and weight of fortifications is a clear reminder that

Panorama of empire: Sea view from Al Jalali Fort in Old Muscat, from where the Portuguese surveyed what they sought to command in sixteenth-century Oman.

the inhabitants here were not merely nomadic herders living in tents, but also some of the finest architects and engineers of any time.

These fortifications are relics of European expansionism, structures that are still bringing benefit to their formerly colonized lands, as so often, they have become key attractions for tourism. This is so not only because of the intrinsic value of the heritage that these structures embody and represent, but also because of the links to the cultures from which today's visitors come, most obviously Portuguese tourists (of which there cannot be too many, given the size of Portugal) visiting the places that their ancestors used to rule and lord over. This unforeseen benefit of imperial legacy is an irony witnessed in many a former colony of a European power elsewhere around the world. Imagine how much more impoverished travel would be without these vestiges of colonization.

Much as Saif dons his cultural heritage not too heavily, Oman wears the legacy of imperialism lightly too. Of course, relics such as the forts and other ancient structures still stand. But there is much less that stands out in modern-day Oman of the European episodes of its history, compared with other formerly colonized lands that may still display legacy elements such as language or public statues. This is another nation brand attribute of Oman—its assured national identity, perfectly comfortable in its coexistence with the foreign influences of its past, as much as the external influences of the future that it has so warmly opened the doors of its world to. Continuing to leave these doors open will help see Oman through Covid-19 and beyond.

Madagascar

Microcosm of survival

The haiku poem and ink sketch in this haiga artwork liken the nation of Madagascar to a forest reserve that can bloom luxuriantly, as if from the seed of a baobab tree. The main country brand attributes—the fruit—include its powers of sustaining itself despite the challenges of the wild, and to spark joy in visitors who encounter nature's abundant wonders on this island, coexisting with many unique species such as the lemurs.

MADAGASCAR
Natural reserve
like a baobab seed,
resilience, delight, flower
wherever lemurs cede
許木松
2020

Madagascar's 'mascot': The ring-tailed lemur, its signature tail with black-and-white rings spread out on full display in this photo, is the most recognizable of the more than 100 species and sub-species of lemurs found in the wild nowhere else but on this island.

Transit to Madagascar:

'Boleh' in Addis Ababa, Ethiopia

Getting there, as they say, is half the fun. But sometimes, it can also be nothing short of nerve-racking. Getting to Madagascar, in itself, can be something of a mini-adventure. From Muscat in Oman, we returned to Dubai in the UAE to catch an Ethiopian Airlines flight into Addis Ababa, the capital of Ethiopia, on transit. This plane was scheduled to land at 7.55 a.m. From there, the connecting flight to Antananarivo, capital of Madagascar, was set for 8.50 a.m. This meant that we had a mere fifty-five-minute interval between the two flights, which would require a speedy passage through what would, in all probability, be a crowded terminal. Could we make it in time? All kinds of possibilities flashed through our minds. The worst-case scenario was to miss the flight to Madagascar, and so, have the rest of our air tickets forfeited as punishment by the airline for a no-show.

Leap of faith: Addis Ababa airport is often too crowded to walk fast through, but one must believe in being able to make very tight flight connections.

Before we left Singapore, the assurances by email from the airline booking staff were that fifty-five minutes to catch the next flight would be more than enough. There seemed to be no other choice but to cross one's fingers and go with the flow on arrival. I pictured having to sprint some of the way, possibly even across the tarmac in the rain if there was no aerobridge available. There was one homonymic coincidence that I looked to, to give a bit of relief, and maybe an iota of added confidence as well. The international airport at Addis Ababa is named after the sub-city where it is located—Bole—which sounds like 'boleh' in Malay (literally, 'can', meaning 'can do' or 'it's possible').

As it turned out, the flight from Dubai was smooth and without incident and landed almost on the dot. So far, so good. The next challenge was whether the passage to the next departure gate would be blocked by the expected crowds, the inevitable security screening and possibly excruciatingly slow queues. Upon entering the terminal building, the first sight was not encouraging. The place was chock-a-block with people in visible agitation, with their no less unsettled luggage, mostly jostling in snaking queues leading right up to the glass doors before the departure gates.

An airport official came up to the gate in the terminal building on arrival and told the disembarking passengers with connecting flights to step aside and wait for him. Most of us complied. A handful started remonstrating, stating their fears that they would not get to their gates on time. A few minutes elapsed as we waited for the rest of the disembarking passengers to make their way through the plane. One Caucasian passenger huffed a little, then gave up and stormed off to try to make his own way through the crowds. That would turn out to be a serious mistake for him. What happened next was that the official beckoned to those going to Madagascar to follow him. Because we were with him, we could overtake many other people standing in line to get through to our gate. This was the only way we could have made the connection in time. Ordinarily, no one would have let us cut queue.

Here was one instance in which patience and adhering to an improvised system is the best option—not quite 'going with the flow'; more like 'going with the know', that is, taking the advice of those who know where to go. Addis Ababa airport is a microcosm of human

Over Addis Ababa: Aerial view from the flight from Dubai into Bole International Airport.

adaptability. When the larger operating system—the one that includes scheduling of flight times and coordinates the layout of connecting departure gates—has reached its limits, then additional human intervention is required to enable business to be sustained. The lone passenger who had stormed off to go his own way was more likely to come up against a non-negotiable barrier somewhere along the line. He would be, as it were, as likely to succeed as a lone activist against an authoritarian regime.

Madagascar was to be another kind of microcosm, one that exemplifies so many marvels of survival.

What is also needed is for passengers to have faith that things will work out somehow. They require some measure of flexibility to put aside their expectations drawn from the norms of airports in other parts of the world. This is one way for an airport like this to survive in such a competitive global industry. Now, with Covid-19, Bole Airport must be much, much quieter, a stark contrast to what we witnessed that day. This episode in Addis Ababa of airport industry survival, and of passenger adaptation, served, in its own way, as a kind of appetizer for the wondrous universe of natural adaptation that Madagascar was about to open up for us. In what might appear at first sight as a massive wilderness, there is order and method in the ecosystems that have found their own ways to thrive, improvising and wending their way through dense forest, just as we had navigated a packed airport terminal to reach our connecting flight.

Lemurs and other marvels of survival

Madagascar was to be another kind of microcosm, one that exemplifies so many marvels of survival. Madagascar nicknames itself Grande Ile (French for 'Big Island'), and deservedly so, as it is the fourth largest island on Earth (after Greenland, New Guinea and Borneo). But it remains relatively little-known compared with other top nature spots.

Pop culture portrayals of the lemur have catapulted it to the international 'A list' of icons of the animal universe on screen. Before the 2005 DreamWorks animated movie *Madagascar*, this island definitely would have had a much lower brand recall globally. The movie and the ring-tailed lemur, in particular, have done more than anything else to bring this African nation to modern global attention. The popularity of the basic story and cast has continued to burgeon since the first release, ensuring it a permanent place on the video playlist everywhere through the *Madagascar* franchise in film and on TV, including the Netflix prequel series *All Hail King Julien*.

To see in the wild the ring-tailed lemur that inspired the character King Julien XIII in the movie *Madagascar*, there are a few favoured spots in Madagascar. One of the most accessible is Berenty Reserve along the Mandrare River, set in the semi-arid spiny forest ecoregion of the far south of the main island. The collective noun for a group of lemurs is—yes, indeed—'a conspiracy'. Watching their natural behaviour, it is not surprising that linguists picked this word. The lemurs possess boundless energy, always look as if they are whispering to each other and do indeed

A conspiracy of lemurs: Sunning themselves, as if in a loose formation, in the early morning light at Berenty Reserve.

On bended knee: For visitors' amusement, ring-tailed lemurs at Vakona Forest in Andasibe have been conditioned to clamber and clutch onto any perchable part of your body, even right on top of your head if you allow them to.

Floral link with Southeast Asia: A hibiscus plant in Andasibe-Mantadia National Park.

appear to be colluding non-stop to locate the next meal or to investigate the next sight of interest within their very wide fields of vision.

Lemurs typically move around together in a conspiracy, and they are constantly communicating with each other on where to explore, or what to forage for, next. They are often spotted plucking berries from trees but would be just as happy nicking bits of human breakfast from an unattended forest lodge table, if allowed to. Lemurs are a target for the one major threat known as man, through poaching for the bushmeat and pet trade. But human threats aside, their survival appears to be relatively easier, and lemurs can be so relaxed generally because predators such as the fossa, an endemic wild cat that looks like a small cougar with dark brown fur, have had it even worse from environmental degradation, habitat loss and poaching. In places like this reserve, thankfully, lemurs can roam almost entirely unmolested.

Generally, the animals of Madagascar do not instinctively fear humans, thanks to the island's centuries of isolation from human encroachment and the mostly unthreatening encounters with humans in today's ecotourism. One of the most awesome yet amusing sights in nature must be to watch, from just a few metres away, a sub-species of lemur called Verreaux's sifaka prance across a dangerous area such as an open track. First, they stop at a tree close to the forest path to check that

Africa's Galapagos Islands: Madagascar hosts many endemic species, including unique mammals such as (clockwise from top) Verreaux's sifaka, the fossa and the indri.

the way is clear, somehow failing to spot the humans watching close by with their cameras clicking, before dancing, one by one, to the other side. The one-by-one method of crossing is what soldiers would do in the jungle, but these lemurs' posture and movement is a cross between what one would see on a ballet stage and in a circus ring. We had the good fortune to watch this sifaka dance a few times. The videos we captured will provide many future hours of happy reminiscence.

Animals that have been victims of poaching would probably behave differently, but lemurs in places like Berenty Reserve will quite happily allow photographers to come right up to almost an outstretched leg's length away. At Vakona Forest in Andasibe in the central-eastern part of Madagascar, lemurs have been conditioned by local guides, lured with snacks to come right up to tourists and—for those who are comfortable with all species of animals—even pose on your shoulder for a smartphone wefie.

There are many other types of lemur, including the indri, also called the babakoto, which is one of the largest living lemurs and is noted for its distinctively plaintive cry, which has a melodic touch of melancholy. It is one of the most moving sounds in nature. The etymology of its name is quite cute—'indri' simply means 'there it is!' in the native Malagasy language. The species got its name from local people shouting 'indri, indri!' when pointing out the animal, just like excited tourists and guides do to this day. Our guides also did so, when we were there.

Madagascar is, in some ways, like the Galapagos Islands of Ecuador. Being similarly cut off from any other large land mass, Madagascar hosts a biodiversity that could thrive and evolve magnificently. Of the more than 200 mammal species, 95 per cent are endemic, found in the wild nowhere else.

Ancient links to Southeast Asia

In the movie *Madagascar*, the cartoon character King Julien XIII, leader of the ring-tailed lemurs, was reportedly originally meant to be just a 'two-line' part. But then, the English actor Sacha Baron Cohen (best known for satirical roles such as Borat), who was auditioning for a voiceover role, improvised eight minutes of dialogue in an Indian

People of Madagascar: A young woman on the road to Berenty and children in a village on the Anjajavy coast show a blend of African and Asian features, reflecting an ethnicity believed to have originated from Indonesia from 200 BCE.

accent, and showed the comic value of secondary players in the movie. Baron Cohen said in a media interview that the Indian accent he used for this lemur cartoon character is based on his Sri Lankan lawyer in real life. As it happens, there *is* some geographical and historical basis for making the Madagascar–India connection.

Geographically, Madagascar can be said to be a 'neighbour' of India across a wide expanse of water, with the Indian Ocean flanking the island's eastern coastline and the Indian subcontinent some 5,000 km to the northeast. According to the theory of plate tectonics, the two pieces of land were once conjoined, with Madagascar having broken off from the land mass of India some eighty-eight million years ago. Today, people of Indian descent make up just about 1 per cent of Madagascar's 2019 population of some twenty-seven million, but many of them are businessmen, and are believed to command as much as 60 per cent of the economy.

The ethnic make-up of Madagascar's population has even closer and deeper links to Southeast Asia. Aeons ago, when Gondwana, the supercontinent that formed most of the southern hemisphere, began to fracture, the waters of the Indian Ocean to the east and the Mozambique Channel to the west flowed across to isolate Madagascar, leaving it 400 km from the nearest human civilization on the coast of Africa.

Humans did not set foot on this island until between 200 BCE and 500 CE, when the first settlers arrived in outrigger canoes from the Sunda islands, some 9,000 km away in what is today the Riau Archipelago, including Indonesia, Brunei, East Timor, Sabah and Sarawak. These intrepid sailors were relatives of a larger group called the Austronesian peoples, whose own maritime migration spanned the seas to cover also the area stretching from Taiwan to Oceania.

It is quite marvellous to imagine how these ancient explorers could have dared to venture out in such flimsy seacraft across such huge distances over choppy waters, with no modern navigational aids and perhaps relying only on the sun and the stars for directional guidance. What is also amazing is that they were able to overcome hunger, seasickness and, most of all, self-doubt, before settling on an uninhabited island with strange and dangerous animals to try to eke out a new life.

The first human settlers arrived in outrigger canoes from the Sunda islands, some 9,000 km away in what is today the Riau Archipelago.

Madagascar's ancient linkages to Southeast Asia are truly quite intriguing. Malagasy, the national language of Madagascar and the main indigenous tongue spoken there, is the only Austronesian language used in the whole continent of Africa. Its closest linguistic relative is the East Barito branch of Malayo-Polynesian, which includes about a dozen languages originating in south-eastern Borneo, today part of Indonesia. For example, the Malagasy word tànjona is believed to be a loanword from Malay, in which the word tanjong means 'cape'. This move from Asia to Africa remains one of the greatest mysteries of international migration. It is also a wondrous tale of survival and adaptation.

Another cultural link to Southeast Asia is in how rice became the staple food for all three typical Malagasy meals of the day—as porridge for breakfast, and with meat and vegetables for lunch and dinner. On average, each person in Madagascar reportedly eats some 120 kg of rice per capita per year, with only Vietnamese people eating more, at 140 kg annually. By contrast, rice consumption in China is only around 70 kg per person per year.

Dietary link to Southeast Asia: Rice is grown all across Madagascar, seen in these paddy fields on the outskirts of the capital Antananarivo and from the air on a flight to Anjajavy.

Paddy is the number one crop in Madagascar, and rice is so representative of, and yet ironically so common to, Madagascar, as alluded to in the poem titled 'The Common Lover's Song' by the Malagasy poet Flavien Ranaivo in the book *Bending The Bow: An Anthology Of African Love Poetry* (2009). In these lines, rice is used as a metaphor for something so commonplace that it is easily taken for granted, the shared fate and pain of all victims of unrequited love, that, 'once swallowed, you think no more of':

> Do not love me, cousin,
> like a shadow
> for shadows vanish with the evening
> and I would keep you with me
> all night long;
> . . .
> or like rice
> for once swallowed
> you think no more of it

Madagascar, indeed, is a microcosm of the world, home to some eighteen ethnic groups. The island's original inhabitants from Austronesia and Southeast Africa were later joined by settlers from East Africa: Bantu Africans and others who, together, through centuries of inter-marriage, make up today's modern Malagasy population. As a trading stop off the African coast, Madagascar also absorbed Arab influences from seafarers from the Middle East, and, at various times, the island was also fought over by the British and the French. French colonization—asserted at times by military suppression of uprisings and by sending Malagasy royalty into exile—lasted from the first invasion in 1894 until independence in 1960. The legacy of colonialism is also another cultural aspect that Madagascar has had to adapt to in order to survive.

Colonial legacy: Still a 'French window' to the world

Madagascar has many legacies from its colonial heritage, and French imperialism has left some long shadows, as if in the wake of a glorious sunset. The first French settlement in 1643 was on Fort Dauphin, the former French name now replaced by Tolagnaro in Malagasy. The old fortifications became relics long ago. Even more recent French establishments, such as the old Miramar Hotel near a cliffside where we watched one particularly spectacular dusk, had fallen into disrepair, the faded red paint of the hotel name just about legible on the crumbling stone boundary walls.

Today, Madagascar remains a Francophone country. French is spoken amongst the educated population, a necessary bridge to the developed world for the elite of society, who thus gain access to opportunities not available to the Malagasy-speaking majority of the population. In a remote village school near Anjajavy that we visited, French is used to teach subjects including mathematics. No proper roads lead to Anjajavy, a peninsula accessible only by boat or plane, but a school like this is a 'French window' to the world for children from villages stretching for many kilometres around.

Madagascar has many legacies from its colonial heritage, and French imperialism has left some long shadows, as if in the wake of a glorious sunset.

Our Malagasy guide was called Olivier, a typical French name, part of the common cultural legacy of many citizens of African nations that once were under French occupation, such as the Arsenal footballer Pierre-Emerick Aubameyang of Gabon, a French colony from 1903 until 1960, or Didier Drogba of Chelsea, who is from the Ivory Coast, a French colony from 1893 until 1960. Our guide Olivier was fluent in French and English, naturally for the tourism trade. Also naturally, he was keen to learn from us a few conversational phrases of Mandarin beyond 'Ni hao ma?' ('How are you?'), to facilitate his rapport with tourists from China, a growing visitor segment here, as in many parts of the world in the pre-Covid-19 world.

Like Tahiti in French Polynesia in the Pacific Ocean, an overseas territory of France, the overall framework of Madagascar's economy has two main 'zones'. One zone includes the high-end luxury resorts and ecotourism in choice spots (accessible only by boat or plane and

Legacy of colonialism: Mathematics classes are taught in French (see the chalk writing on the board, still streaked white with chalk dust from a lack of water) in this village school near Anjajavy.

sustained by supporting industry sectors with the highest-paying jobs). The other zone covers everything else, including a capital city that could do with more maintenance and development, and the rest of the terrain, which is a kind of huge 'flyover' country, where subsistence agriculture is the only livelihood for almost three-quarters of the population.

Across such terrain, the theoretical 'trickle-down' of economic benefits is a bit difficult when the trickling must try to drip through and across an area of almost 590,000 sq km, and when the flow might already face some hindrances and challenges near the top, let alone anywhere further down the line. Unfortunately for Madagascar, like many a former European colony, throwing off the yoke of colonization did not bring enduring universal bliss. Political survival has been difficult ever since self-government was on the agenda. In the wake of the retreat of French imperialism across Africa, Madagascar's First Republic (1960–1972) was established as a democratic system modelled on that of its former colonial masters.

The new nation was tiptoeing forward from what was the only mode of governance it had ever known up till then. But the lingering

Royalty in ruins: The Rova (palace) of Antananarivo, a reminder of Madagascar's ancient royal past, is situated on a hill and is visible from almost anywhere in the capital.

economic and cultural dependence on France sparked a backlash of popular movements amongst farmers and students from energies that had been suppressed before, and this led to a socialist democratic republic from 1975 to 1992, when external assistance was sought from states aligned with the Soviet Union. However, this did not bring higher standards of living but more social unrest instead. The 1992 multiparty elections ushered in the democratic Third Republic (1992–2009), and things have certainly stabilized and improved since then. However, the potential for political crisis and popular movements has continued to bubble underneath the surface calm ever since.

One of the consequences of the stark contrast between the relatively less-developed parts of the inner city in the capital and the lavishness of the outlying luxury resorts is that tour operators—just like in the Maldives or Fiji—typically try to dissuade tourists from spending any more time in the capital city than is necessary to get on to the transport connection—whether by road, air or sea—to travel on to the tourist spots that cater to international tourist standards. In the case of the capital Antananarivo (called 'Tana' for short by the locals), there are definitely quite a few attractions worth seeing, even though, for safety concerns, independent exploration by foreigners must be done with care.

The French language is a 'French window' to the world.

The two main sights are linked to the ancient monarchy that once ruled Madagascar. Up on a hill overlooking everything else is a royal palace complex called the Rova of Antananarivo, which can be visited on a half-day city tour. It served as the home of the sovereigns of the Kingdom of Imerina in the seventeenth and eighteenth centuries, as well as of the rulers of the Kingdom of Madagascar in the nineteenth century.

The other main sight is the 'counterpart' of the Rova, in the nearby fortified village of Ambohimanga, which served as the spiritual seat of the kingdom in contrast to the political significance of the Rova. If Antananarivo was where the kings lived in their pomp, Ambohimanga is where they are now remembered when they died. This UNESCO World Heritage Site in the central highlands was a spiritual centre for the Merina people of the Kingdom of Imerina (circa 1540–1897).

Ambohimanga was home to numerous kings, including Andrianampoinimerina, who ruled from 1787 to 1810 and undertook the ultimately successful initiative to unite the island of Madagascar. This was also the designated site of many sacred royal rituals, and the

Royal excursion: A school excursion group makes its way to the ancient burial sites of Ambohimanga, dotted with many poinsettia trees, a national floral emblem of Madagascar.

hill and the village that crowns it remain sites of pilgrimage to this day—reminders of a heritage that did not endure. In the same way, the Rova palace did not survive either—it was burnt in a fire in 1995, which our guide says is suspected to have been an act of arson, and this jettisoned the bid to have it inscribed as a UNESCO World Heritage Site.

It is a pity that this rich heritage of Madagascar is not better preserved. This untapped reservoir of historical richness makes more poignant the aspect of 'hope' that the green colour of the poinsettia plant represents as part of the national floral emblem. This aligns with the nation brand attribute of what might be called 'fertile potential', which is very much at the heart of Madagascar as a microcosm of survival and adaptation. Such fertility is truly lush in the forest, which is where we went next.

Baobabs and reptiles: Symbols of resilience

Survival in Madagascar is represented by a few metaphors. From the botanical world, there is no better ambassador than the baobab, with its awe-inspiring hardiness. The baobab (genus: Adansonia) is not unique to Madagascar, but this island is its most glorious home in the world, definitely in terms of global brand awareness. Of the nine baobab species acknowledged by botanists as of 2018, six are native to Madagascar, two are native to mainland Africa and the Arabian Peninsula, and one is native to Australia.

One tagline of the National Tourism Office of Madagascar—'Treasure island'—works because it has the two key qualities of such a catchphrase: It has instant recall, at least to an audience broadly familiar with Western culture, within which this tagline alludes to the title of the classic 1883 adventure novel by the nineteenth century Scottish writer Robert Louis Stevenson. It also contains a double meaning, prompting mental pictures of the treasures of nature that can be found on this island, the unique fauna such as lemurs and flora such as baobabs.

The baobab, called 'the tree of life' in Madagascar, is one of the most impressive trees on Earth, in more ways than one. It can reach a height of up to 30 metres and have a trunk diameter of over 10 metres. To endure harsh drought conditions, baobabs store water in their trunks up to 120,000 litres. In this way, the baobab is a bastion of self-reliance,

Stand with me: The baobab, built to withstand severe drought, is a metaphor of survival for Madagascar.

the botanical equivalent of a camel, adapting to survive the most arid conditions.

Baobabs are deciduous, shedding their leaves during the dry season. And because Madagascar is an island, unlike its landlocked African neighbour nations that also have baobabs, here there are also coastal baobabs, which are an utterly impressive sight, some of which can be seen only from the water by boat. Baobabs can thrive even in the most seemingly inhospitable environments—for example, out of dry, hard rock that would seem incapable of supporting any animate life, like the tsingy limestone of the Anjajavy Forest. By the coast, baobabs seem to perform prodigious acts of balance as well, leaning out over the salty water as if reaching for more hydration, that is so near and yet so far.

The most ancient baobabs have been estimated to be more than 2,000 years old, but in this twenty-first century, the oldest and largest baobabs became parched and dehydrated, and have begun to die, possibly due to climate change, from an unprecedented combination of longer and harsher periods of drought and more scorching heat than ever before. That they are called 'upside-down trees' (because of the appearance of their branches resembling tree roots but reaching up into the sky, especially without leaves in the dry season) is appropriate, given the way they also turn most expectations upside-down in their quest for survival.

Baobabs can thrive even in the most seemingly inhospitable environments.

Baobabs have an impressive heft and shape. In any context, they would grab attention, such as when transported halfway around the world into a biodome at the Flower Dome of the Gardens by the Bay in Singapore, futuristic public gardens that replicate Mediterranean and montane ecosystems to display plants from these regions. But these trees look especially stunning when standing alone in the midst of a surrounding landscape that is comparatively flat, which is where many can be found in Madagascar.

Another symbol of survival for Madagascar is the adaptability of its reptilian inhabitants. About half the world's chameleons are unique to Madagascar. Currently, almost seventy species have been identified, and new ones are still being discovered. The Belalanda chameleon (Furcifer belalandaensis) is endemic not only to Madagascar but to just one district—Belalanda, a town and rural commune in Toliara city, in the region of Atsimo-Andrefana, in south-west Madagascar.

Sometimes, you have to trek for hours to look for special chameleons; sometimes they just appear right next to you. Our sharp-eyed driver pointed one out right next to our car, on the road out from Tolagnaro

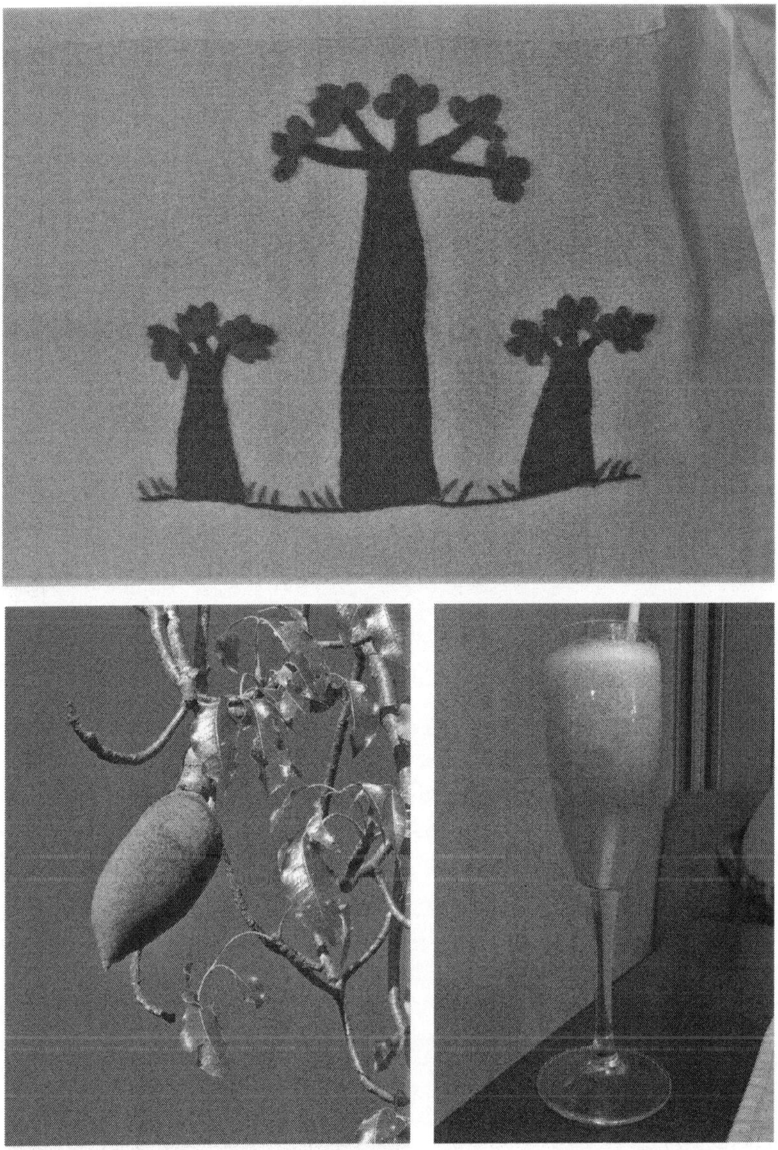

Icon of nature: Apart from being an impressive sight, the baobab can be enjoyed in various other ways—in celebratory embroidery stitched lovingly by village women, or in its surprisingly substantial juice, drawn from its plump fruit, as refreshment in a restaurant.

Right by the water: As an island, Madagascar offers uniquely awe-inspiring sights of coastal baobabs, as on this island near Anjajavy.

Airport, on the way to Berenty Reserve, soon after we had arrived on an internal flight from the capital Antananarivo. Internal flights are the best way to get to different spots across this vast country, not only because of the great distances but also because of the condition of some of the roads built by the French that have not been maintained well enough ever since. Spotting wildlife was a splendid diversion on the transfer, on roads that were so uneven that the drive took almost four hours even though the distance was just around 80 km. The green of this chameleon's skin perfectly matched the leaves around it. Only the thin, speckled white stripe along its side gave away its presence, but even then, only to the most expert spotters.

The Belalanda chameleon is one of only five critically endangered reptiles. This is because it has a very small range of habitat, being restricted to an area of 4 sq km. Its original habitat was gallery forest, but this has all been cleared, and it is now found usually in the canopies of the non-indigenous trees that have been planted and in the few remaining mature trees. Its main threat is, once again, from man's activities—in this case, from the significant amount of timber logging for the manufacture of charcoal in this area.

More than surviving: The forest-camouflaged Belalanda chameleon (left), endemic to just one town, and the more numerous Malagasy giant chameleon (above), clambering on a beach boardwalk.

Madagascar's national animal: Lemur or zebu?

Visiting Antananarivo gave me the opportunity to meet my friend Fara, a government official who works in foreign affairs and development and looks after Madagascar's nation brand. We had met earlier at an international place branding conference in London. In our meeting in Antananarivo, which also included her colleague Edmond from the same department, we chatted about brand Madagascar. Snippets of this conversation were published a few days later on her department's social media platforms. It was at this meeting that we enjoyed the ichigo ichie moment of tasting a glass of baobab juice, with the milkiness of barley, half the sweetness of longan and just a hint of the sourness of soursop.

I noted that the image of Madagascar remains that of a country particularly rich in terms of biodiversity. This is a competitive advantage that the Big Island must exploit. More could be done here to foster awareness of the wild beauty of this ecosystem. Lemurs are well-known internationally, thanks to movies and pop culture, but there are so many other species of fauna and flora that also deserve the limelight. There is

Global star: The ring-tailed lemur (seen here at the Singapore Zoo) is the internationally well-known logo of the Madagascar National Parks.

much potential to be tapped in the cultural diversity and know-how of the Malagasy people, to bridge any existing awareness gaps.

I also suggested that it is not prudent to include all aspects of a country in a catch-all brand. Instead, placemaking agencies in Madagascar should define, together with all national stakeholders, what are the main markers of Madagascar's brand identity. Madagascar could make itself better known for its economic potential, its people and its culture, along with its biodiversity. To do this, the coordination of all relevant actors in the private and people sectors is essential. Perhaps Madagascar's X factor is that, as an island country, it can benefit from its natural isolation to consolidate its image of an independent country with extraordinary natural richness.

In a separate meeting with Tahiry, another Madagascar official in economic development whom I had also met at the London conference, one of the topics we chatted about was the country's national animal. Hitherto, the lemur has been Madagascar's best-known wildlife representative. This is consistent with international practice, for aesthetic more than practical reasons. What makes an animal an icon is that it should, above all, be charismatic, never mind if it is not easily accessible. For instance, the panda represents the nature aspect of China, even though only a minority of people will get the chance to come up close to a panda in real life. Despite the lemur's greater 'fame', Tahiry believes the zebu should be promoted and come to serve as a national metaphor. Amongst other factors, the zebu is a traditional symbol of family wealth, strength, self-reliance and resilience.

In Malagasy culture, the zebu has always been a status symbol, like other primary livestock in most agricultural societies. This variety of humped cattle can also evoke ideas such as status and self-sufficiency. My view is that both animals can help, from inside out. Perhaps the zebu could play a larger role in the internal brand-building of Madagascar, as a focus of inspiration for the people, in tandem with complementary efforts of economic development. Meanwhile, the lemur could continue to be promoted externally as the country's external wildlife symbol that is best-known internationally, while leaving its internal standing on 'auto-pilot', as it were, as it is already fully established in the popular imagination. This dual dimension of nation brand-building—to invest

Status symbol: A zebu seen off the road to Berenty Reserve, and represented as a motif on a lamba shawl woven from traditional materials, which is presented to visitors as a gesture of friendship and respect.

in internal country brand-building while driving the external brand promotion—is something many countries would not even contemplate, since the internal dimension of nation brand-building is generally neglected because it is not properly understood to begin with.

For Madagascar, once again, its relative isolation—both in terms of geography as well as in its networks of connections with the rest of the world—is both a unique attraction as well as a significant challenge for both internal and external brand-building. Being able to draw on the gravitas of a zebu and the global fame and guile of a lemur, what is without question is the immense potential for the Big Island's nation brand to go on to much bigger things in the future.

Splendid isolation: A ghost crab on an Anjajavy beach catches the last glints of sunset.

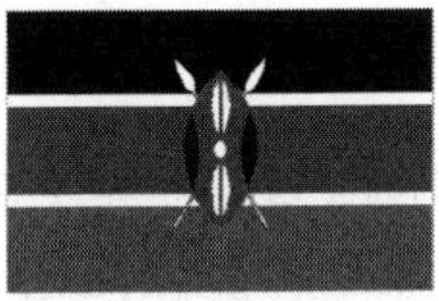

Kenya

Humility and wonder

The haiku poem and ink sketch in this haiga artwork preserve a defining sight in Kenya and how these images showcase the chief attributes of the country brand—the magnificent Mount Kilimanjaro, playing hide-and-seek in the cloudy rainy season, appears at dusk in unison with a double rainbow. A deep humility is evoked by an appreciation of how small and inconsequential man is in front of the awesome scale and beauty of nature.

KENYA

Divine view

wet season's thick snow
caps what's already blessed with
this double rainbow

許木松
2020

Majesty of nature: The wilderness of Kenya inspires reflection on the relative insignificance of man.

Into Africa

'The Earth was made round so that we would not see too far down the road,' says the Danish writer Isak Dinesen, played by Meryl Streep in the 1985 movie *Out Of Africa*. In evoking the romance of what was once known as the 'Dark Continent', the impact of this drama in shaping the celluloid world's sense of Africa was huge—as monumental, one could say, as Mount Kilimanjaro is to the nation of Kenya.

In reality, visibility in general is usually quite good across Africa's savannah grassland. Even in the warmer months as when we visited, with the grass growing long, one can often see far into the distance, further than at ground level, by comparison, in the major urban centres from which most safari explorers commence their adventures into Africa. And, if one is equipped with a decent pair of binoculars, one can spot not just the considerable heft of, say, an elephant on the horizon or ridgeline, but also the slenderest, most elusive cheetah camouflaged amongst the long grass, or even a furtive hyena slinking away through the undergrowth.

Most people associate Kenya with vastness, of an expanse of nature that remains not fully explored even centuries after the first encroachment of man.

Talking about seeing far down the road, most people associate Kenya with vastness, of an expanse of nature that remains not fully explored even centuries after the first encroachment of man. As with most other aspects of life, for nation brands, size usually matters, except for those countries where the size of the brand exceeds the physical extent. It is a sign of Kenya's enormous country brand equity that most people would be surprised to be told that, in the list of countries by land area, Kenya, which they might imagine to be a huge country, actually ranks just behind Madagascar, where we had just visited, which most people would guess is a small island before looking at the map to check.

Western observers have had the most impact on shaping the world's impressions of Kenya, and while this is generally true of almost everything else, Kenya does hold an exceptional place in the perceptions of earlier

European encounters with the continent. Isak Dinesen was the pen name of Baroness Karen von Blixen-Finecke, whose memoir published in 1937 gave not only one colonist's perspective, but also a window into the changes to Kenya brought by British and other European farmers, who became wealthy farming coffee and tea in Kenya's interior central highlands during the early part of the twentieth century. By the 1930s, approximately 30,000 white settlers were living in the area and had gained a political voice because of their contributions to the market economy.

Making a mountain: Enjoying a rare clear view of Kilimanjaro's snow-capped summit, and celebrating it on fabric for the tourist market.

Unlike Dinesen, Ernest Hemingway's literary references to Kenya are somewhat less rosy, for example, in *The Snows Of Kilimanjaro,* a short story first published in *Esquire* magazine in 1936 and adapted into a film in 1952 starring Gregory Peck and Ava Gardner. 'Love is a dunghill, and I'm the cock that gets on it to crow,' Hemingway's protagonist Harry says, as he contemplates failure and disappointment in his life, in this exploration of isolation and alienation. In this story, Mount Kilimanjaro can be read as a symbol of truth, idealism and purity. In another of Hemingway's symbols of the purity of nature, a leopard said to have been found dead on the mountain had perished in a high, clean, well-lighted place, in contrast to the desolation of human existence.

The imagery of Kenya described by Dinesen and Hemingway evokes a sense of perspective of the place of man against the vastness of nature. Mount Kilimanjaro is the highest peak in Africa at 5,895 metres. Like Japan's Mount Fuji, it is not always available for admiration, to say the least. On many days during the wet season, the peak is completely

Four of the Big Five: Clockwise from top left—a leopard, lioness with a buffalo kill, buffalo and elephant.

shrouded with clouds, or the mountaintop is obscured by thick mist. On other days, even when the sky is otherwise clear, a few swirling clouds insist on dancing around the top, as if intentionally teasing and denying a full view of the crest. In playfulness, a solitary cloud would sometimes loiter precariously around the edge of the summit.

We flew into Nairobi's Jomo Kenyatta International Airport from Antananarivo via Addis Ababa, and this time, with a longer interval between flights, the transit through Ethiopia was uneventful. We met our guide Joseph the next morning. The day we saw the peak of Kilimanjaro, he told us how fortunate we were to be the first visitors this season in early summer to have been treated to a clear view after weeks of rainy weather. The ichigo ichie offerings of nature were to become even more fulsome. The next day, we were doubly blessed to see a full rainbow, and a double rainbow, at that. Enjoying the sight of a rainbow across the savannah is a treat of the wet season that is seldom possible when the weather turns dry, for rainbows form only after rain. So much of the wonder of travel is cherishing the little mercies and gifts that can fall one's way at any time, on any day.

Of all the joys of travel, there are few as majestic as an African safari. This is the completely sound basis for the 'magical Kenya' slogan and imagery of the tourism sub-brand of Kenya's country brand. For this sub-brand, the crown jewel is Kenya's Maasai Mara National Reserve, a 1,510 sq km stage for some of the most stunning wildlife on Earth. This is also the venue for one of the great spectacles of nature every July and August when over a million wildebeest and thousands of herd animals migrate here across the Mara River in search of lush grass, only to turn south for greener pastures in October and November, and then to repeat this cycle every year. Our itinerary did not coincide with the migration season, but nonetheless, we were determined to savour every moment in this sanctuary.

Here in May, outside the migration season, it is often rainy, with showers almost every day. The downpour the previous day had softened the mud on the ground and the tyres of the safari jeeps had carved deep, jagged tracks through the main paths leading to the safari camps and across to other destinations such as the entry and exit gates of the national park. The camp we were staying in was on the banks of the

Fracture of Earth: The Great Rift Valley stretches 9,600 km from the Red Sea on the northern Egyptian coast down to the south-eastern shores of Mozambique.

Mara River, which gave us river views from the terrace of our tent and ample opportunities to spot wildlife, ranging from hippos on riverbanks in the distance to groups of banded mongoose scurrying around just below and around our tent.

One consequence of our camp being located further into the national park was that we had more time, at least twice every day—after breakfast and just before dinner—on the 'roller-coaster' rides through the thick and slimy mud to and from camp. Some days, we also returned to camp for lunch. Once, we experienced the full extent of the phrase 'stuck in a rut', when a wheel on our jeep got jammed in a particularly deep groove in the mud, ground out by the jeeps that had passed this way earlier. Joseph had to park the jeep, step out to look for a rock, put it in front of the wheel and then rev the engine so that the wheel would be able to leverage itself over the rut. Many times, we felt glad to be in a sturdy four-wheel drive, steered by an experienced guide who was always confident and careful.

Lakeside safari: Lake Naivasha is home to many varieties of wildlife, including (anti-clockwise from top) zebras and waterbuck that roam around your accommodation and hippos that wander ashore on some nights, as well as lots of birds such as the majestic African fish eagle.

The wildlife was indeed, as the tourism advertising promised, magical. After the 'appetizer' sights of a few giraffes and gazelles, the opening item of the main course—the first of the 'big five' that we spotted—was a leopard. It was just sauntering along a track in full view, seemingly oblivious to the human activity around it. What a thrill! This in itself was worth the price of admission.

Rhinos, we were told, are a rare sight in these parts, with their numbers sharply dwindled mainly through poaching, so we jettisoned that hope right away. The other three of the 'big five' were spotted shortly, in addition to other species including cheetahs. The lionesses were the most memorable, two of them with three cubs around a dead buffalo that had been killed a day ago but was still providing food that day. The lionesses would probably feed on the carcass for one more day before giving way to the lurking hyenas and loitering vultures. It is always amazing the way that these beautiful creatures can nonchalantly stroll around with their young while so many humans are driving around in jeeps with cameras clicking away.

Of all the joys of travel, there are few as majestic as an African safari. This is the completely sound basis for the 'magical Kenya' imagery of the tourism sub-brand of Kenya's country brand.

Another wonder of nature that visitors to Kenya can enjoy is a view of a portion of the Great Rift Valley. The route from, and back to, Nairobi passes an area that affords a great view. On the way out from Nairobi airport, it was rainy and so cloudy that we could hardly see anything. But on the way back into the capital, the sky was clear-blue through the clouds, revealing the splendour of the valley. The wonder is that this is actually an awesome indication of plate tectonics, of dynamic movements in the Earth's crust—this is where Africa is slowly being pushed and pulled apart by uprising material in the Earth's mantle. Geologists believe that eventually, the rift will crack open deep enough to host a new ocean, separating Eastern Africa from the rest of the continent. The fractured rift will become nothing short of a wondrous sight, and this will be a game changer for the destiny of Africa.

Native spirit: The Maasai people, as depicted on a tourist souvenir in a Nairobi shop, and enlivening the savannah of the Maasai Mara.

Kenya also has much natural beauty beyond the Maasai Mara. One place is Lake Naivasha, northwest of Nairobi, and part of the Great Rift Valley. Naivasha is the hometown of another of my graduate school classmates, Wanja, but unfortunately, she too was out of town so this was another reunion in spirit. Her love for her homeland is deep and warm, and it is not hard to see why. At Lake Naivasha, visitors can live amongst rich wildlife that thrives in and off the water. Each lakeside lodge or resort has its own resident species. These usually include harmless zebras and waterbucks, which graze on their own if left undisturbed. The hippos frolic in the lake most of the day, sometimes out in the open in the water in the brilliant sunshine, sometimes amongst the reeds by the water's edge. On some nights, they venture ashore to roam around the tennis courts and lawns. Security guards will sometimes knock on your room door to call you to go out to take a look. We were called one evening. Dressing hurriedly, we made our way out in time to get at least a few fuzzy photos through the pitch dark of a hippo that had come ashore to roam around the hotel lawns.

From nearby spots, boats take visitors out onto the lake, close to the pelicans perched on tree trunks in the water and swimming hippos amongst the floating hyacinth and other water plants. The boatmen

Tribal economy: Visitors are often surprised to find that Maasai beaded accessories, worn by both women and men, are made from materials imported from Europe and as far as China.

will also throw bait into the water for the African fish eagles to swoop down to catch in their beaks. These handsome birds have probably been conditioned to perform this show to the amazement of boat passengers.

It is not just the animal kingdom that Kenya hosts, but also the people who have lived off this land for centuries—tribes such as the Maasai. They are there to welcome guests right at the entry gate into Maasai Mara National Reserve, usually women who approach safari vehicles to ply their products—necklaces, bangles and other accessories mostly made of multicoloured beads. The Maasai people are one of the best-known native peoples in the world. Aspects of their culture, such as how young men have to prove their manhood by killing a lion, are part of the folklore of Africa that has become familiar across the globe.

The Maasai people originated in South Sudan, their name simply meaning the people who speak Maa, their language. They migrated to their current homeland between southern Kenya and northern Tanzania during the late seventeenth and early eighteenth centuries. They displaced some of the other tribes that had previously settled in the region, while other tribes were assimilated into the Maasai culture.

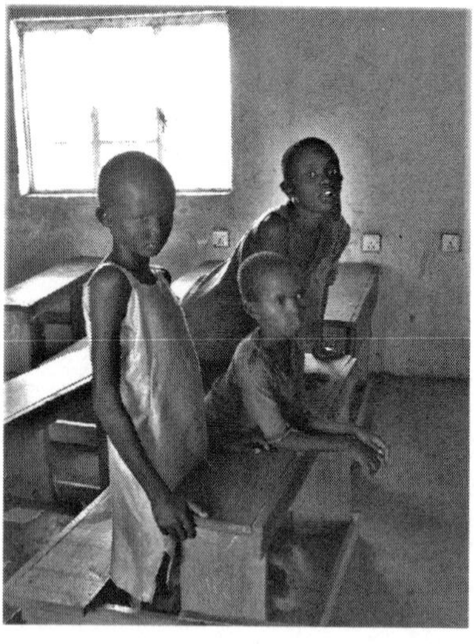

Learning for life: Maasai village children in school and frolicking with their pet animals in the great outdoors.

Their main activity was raising cattle, but the Maasai have also been better known for centuries as fearsome hunters and warriors.

Maasai culture and the management of modernity

The Maasai have been managing modernity for some time. By the mid-nineteenth century, Maasai territory was at its largest, extending over pretty much the entirety of modern-day Kenya and half of Tanzania. The period between 1883 and 1902 was the darkest time in Maasai history. It is known in the Maa language as emutai, meaning 'to wipe out'. It is estimated that up to 60 per cent of the Maasai people lost their lives during this period as a result of smallpox, drought and starvation, after an animal disease known as rinderpest killed almost all of their cattle.

In the early twentieth century, vast sections of the Maasai land were turned into national parks and wildlife reserves. The Maasai, understandably, could not have welcomed this move. Around the same time, the Kenyan government started pressuring the Maasai to give up their traditional semi-nomadic herder existence in favour of farming and a more sedentary life. To this day, a large percentage of the Maasai people have resisted government pressure to settle in permanent homes, distancing themselves from urban areas and continuing to practise a lifestyle that has remained unchanged for centuries. Maasai men take great pride in herding, as cows are their most prized possessions.

Today, modernity has been woven into their lifestyles, although this might not be obvious at first sight. The traditional colourful, often checked, cloth worn by the Maasai may now come from as far as Scotland, which might explain why a tartan pattern seems to be popular. In turn, the Maasai have inspired fashion trends elsewhere. Louis Vuitton's 2012 spring/summer men's collection was inspired by the iconic Maasai shuka, a traditional blanket in fiery red and blue colours.

Maasai jewellery is mostly made of beads, which traditionally were made out of local products—bone, clay, wood, copper or brass—but these natural resources were set aside as soon as glass and ceramic beads became available through trade with Europeans. The Maasai obtained glass beads from as far as India and Murano, Italy, from as long ago as the fifteenth century, when these items were very expensive and used

No worries: Kenyans like to say hakuna matata ('no problem'), the Kenyan equivalent of the Australian expression 'no worries'.

only by royalty. More recently, materials have come from other source countries, places such as the Czech Republic.

Only Maasai women are allowed to make beaded jewellery, and every female is expected to learn this craft. Even the most decorated men wear only what the women in their lives have made for them—their mothers, sisters, wives. Our guide Joseph, who is not from the Maasai tribe, also bought into the Maasai brand literally, as he wore a Maasai bangle on his wrist. The bangle is so tight that he has stopped trying to remove it and now just wears it all the time—possibly for the rest of his life.

A visit to a Maasai village is a highlight of any trip to this area, in addition to the country's vast fauna and flora. However, there is one aspect of such encounters that visitors must accept: the method of marketing to tourists. It is not uncommon for visitors to a Maasai village to be taken to what looks like a small 'market square', where women sit around, each with the beaded items they have made on a mat or

Pride of Africa: Photos on the wall of a shop at a rest stop on the road from Nairobi to Amboseli depict symbols of African pride, including former US president Barack Obama, whose father was Kenyan, and Liverpool Football Club striker Mohamed Salah of Egypt.

cloth in front of them. There is then, shall we say, a focus on assertive encouragement to make a selection for subsequent collective pricing. The leader of the group—the son of this village's chief, on our visit—leads the survey of the market square and the selection of merchandise purchases, and heads the pricing session and negotiation.

This custom is what some might consider an unfortunate development in their livelihood, an invasion of commercialism and the profit motive so deep into the savannah. But this is also something that is understandable, given their situation. This is another dimension, an undercurrent of stress and calculating behaviour that runs beneath, and counter to, the general surface demeanour of Kenyan society, which is characterized by the term hakuna matata ('no problem' in the Maa language) that Kenyans like to say to each other, the Kenyan equivalent of the Australian expression 'no worries'. Once again, pop culture expressions such as those in the movie *The Lion King* have introduced this traditional African catchphrase to younger audiences internationally.

Doing more to showcase the human facets of culture, heritage, entrepreneurship and so on would involve the Kenyan people much more in building the brand of their own homeland—this is the vital ingredient to revitalize any nation brand.

By contrast, the children that one sees in Kenya appear to be shielded, in their youth at least, from this pragmatic aspect of adulthood of having to adapt to the various influences of commercialism. There is still in their eyes a look of ambivalent curiosity about outsiders, if one sees them, say, in a school classroom, as we did on our visit to the village, before they switch their attention just as quickly back to the carefree abandon only possible in childhood, once they are free to scamper off into the outdoors with their friends and pet animals.

The condition of the Maasai also throws light on one aspect of the untapped potential of Kenya's nation brand. The mindshare gained by 'magical Kenya', with wildlife tourism as the mainstay of the tourism sub-brand, also highlights one area of Kenya's country branding that has not received the attention and focus it deserves—the people of Kenya. Doing more to showcase the human facets of culture, heritage, entrepreneurship and so on would help to energize other sectors of the economy. It would also involve the Kenyan people much more in building the brand of their own homeland—this is the vital ingredient to revitalize any nation brand.

At a rest stop on the road out from Nairobi airport, we spotted newspaper clippings put up by the proprietor's family that gave an insight into the continental self-image of the Kenyan people. The pictures in these cuttings pasted on the wall included three of former US president Barack Obama and his family. There was also a photo of the Egyptian striker Mohamed Salah, who plays for Liverpool Football Club in England, currently one of the world's most famous football players hailing from Africa.

The fierce pride of the Maasai is still there somewhere, embedded in the Kenyan cultural DNA. Politicians and other figures of authority in society might disappoint their constituents from time to time with their human shortcomings. But the strength of the Kenyan identity is unshakeable, reinforced as it is by external factors of significant impact.

One of these developments is the Kenyan heritage of former president Obama, further elevating Kenya's place in the consciousness of the rest of the world, reinforcing Kenya's top-of-mind recall as an African nation in comparison with the fifty-three other countries on the continent. Here, promoting national pride is not at odds with humility before the wonders of nature. Indeed, it is this greater awareness of the national identity of Kenya that should form the basis for wanting to protect and preserve what nature has bestowed so generously. For the people of Kenya, there is certainly much more to cherish in themselves.

One more day in Nairobi

The only major blip on this itinerary turned out to be a blessing in disguise in some ways. As things turned out, we had more time to see the heritage of Kenya than we had at first planned, at the end of our visit. On what we thought would be our last night in Kenya, we made our way, as planned, to the Nairobi airport well past midnight to catch our flight to Barcelona, Spain, via Istanbul, Turkey. As we were checking out of our hotel, I glanced at the computer screen above the front desk and noticed that our flight—scheduled to leave at 4.45 a.m.—was not listed. We asked the hotel staff about this, but they shrugged and told us to check at the airport.

On arrival at the airport, it was only when we reached the check-in counter that we realized that our flight had been cancelled. We figured out later that the airline probably did that because there were too few passengers that night, so they moved all the passengers on to the flight the next day, to cut costs. Somehow, we had missed the earlier email or phone messages from the airline, probably due to unreliable Wi-Fi access. Because this cancellation was initiated by the airline, and we had a few days' buffer until the next flight, the resulting one-day delay in our flight schedule would not affect the onward journeys on our round-the-world ticket.

As it turned out, it was a blessing in disguise that our flight was delayed by a whole day. By way of compensation, the airline put us up for one free night at a hotel and provided a car to take us there and back. We were also told that we could order anything we liked from the hotel café.

Making the most of a situation we could do nothing to alter, we appreciated the unexpected change. Having an extra day in Nairobi did not disappoint. It is always an interesting experience to explore a heritage hotel. And with the Sarova Stanley Hotel, there is a lot to savour. Guests can explore the hotel to view its collection of heritage memorabilia that commemorates Kenya's connections with the West.

Kenya is similar to other former colonies in its stance in remembering its colonial heritage, but it also has more substance than most other African nations. The economic importance of Kenya historically—amongst other things, as a port of entry for trade with Europe—had always given it greater importance in the context of British colonialism. This is the economic legacy that Kenya has built till today, enabling it to position itself as one of the 'African lion' economies (along with other better-performing African economies including Ethiopia, Ghana, Tanzania and Uganda) to try to emulate the growth surge of the 'Asian tiger' economies of Singapore, South Korea, Hong Kong and Taiwan in the 1990s.

As far as heritage goes, the Sarova Stanley Hotel is unique—this is where it all started for Queen Elizabeth II, the British monarch who presides over the Commonwealth and all the British territories around the world today. Amongst the displays here is a photo of a luncheon at

Foreign relations: Mementoes of a colonial past and appreciation from the West, in a photo of then-Princess Elizabeth's visit to Kenya in 1952 and a tribute to American writer Ernest Hemingway at the Sarova Stanley Hotel in Nairobi.

the hotel in February 1952, attended by the then-Princess Elizabeth and her husband Philip, then Duke of Edinburgh, at the start of their visit to Kenya. A few days later, they were on safari from the Treetops Hotel in Aberdare National Park in Kenya near the township of Nyeri, when her father, King George VI, died in his sleep in England.

The young princess cut short her trip and returned home immediately to ascend to the throne. She was crowned Queen Elizabeth II at Westminster Abbey in London in 1953. As British hunter and conservationist Jim Corbett (who accompanied the royal couple) put it, she 'went up a tree in Africa a princess and came down a queen'. For all intents and purposes, the queen still is, and most probably will always be, a queen of the Kenyan imagination, even so many decades after Kenya gained independence in 1963.

In these journeys of discovery, part of the fascination lies also in stumbling upon the threads that string together things that happen separately across long periods of time, even centuries.

In these journeys of discovery, part of the fascination lies also in stumbling upon the threads that string together things that happen separately across long periods of time, even centuries. In 1414, the Chinese trader and explorer Zheng He, representing the Ming Dynasty, visited the East African coast on one of his last 'treasure voyages'. He is said to have reached as far as Malindi (near the port of Mombasa) in present-day Kenya, and to have returned to China with such exotica as lions, leopards, camels, ostriches, rhinos, zebras and giraffes. People in China would have been amazed by these animals. The giraffe caused a stir that has been immortalized in art, suggesting that it was considered living evidence of the qilin, a mythical creature representing good fortune that has been compared to a unicorn in Western mythology.

There is nothing mythical, however, about Kenya's connections with China today, although a slice of that sense of scale of the Middle Kingdom might also be in play at this time. As we drove in our safari jeep on highways out of Nairobi, our guide Joseph gave a bright picture of the state of Kenyan economic development today, pointing out port, railway and road infrastructure built by construction companies from China. In

Mutual benefit: Kenya's new economic relationships with investors like China are a bit like the oxpecker-buffalo symbiosis of the Maasai Mara.

a way, these bilateral links are like picking up from a very old story, in Kenya's first major connections with China since the fifteenth century.

China's economic influence today is growing in many other places across the African continent, just as it is in a few other regions we visited on this trip. This reshaping of the global geo-economic landscape is something that may be hastened in the midst of the eventual recovery from the disruptions of Covid-19. Earlier in Madagascar, we bumped into a group of doctors from China who were in Africa for development work. In Kenya, this is part of a larger national development effort, as can be seen with the exterior walls of factories that face the highways painted with billboard-style and -sized slogans such as: 'Here we are building a better Kenya'. The expanding relations between China and Kenya, and also with the rest of the continent, will add to the reconfiguration of a new Africa, one that is in the process of taking shape, further moulded, no doubt, by impetus from the repercussions of Covid-19.

This journey's 'fourteenth country': Tanzania

This journey around the world across thirteen countries, with stopovers in Addis Ababa, Ethiopia, and Istanbul, Turkey, also had a bonus fourteenth country included—Tanzania. One day, on safari in the Maasai Mara National Reserve in Kenya, our al fresco safari lunch had already been made more exciting by the thought that a lion could appear at any moment, without any forewarning. And then, after lunch, our guide Joseph drove us into Tanzania. We went right up to the border and crossed the line into Tanzania's Serengeti National Park. We did not have our passports with us, as these were safely locked in our luggage back at the safari lodge. We did not want lionesses or any other threats to chew up these important documents.

We never stopped to ask Joseph whether tourists were allowed to cross the border; we just assumed it must be something all guides do—and surely, in both directions—given the ease with which the border is crossed, as there is no fence or other barrier whatsoever. To indicate the country border, there are just stone markers erected at spots at certain intervals, presumably all along the boundary, wherever it is practical to put up such markers. We did not linger; only staying long enough to disembark and make sure we physically set foot in Tanzania, if only for a few ichigo ichie moments. Now, looking back, whether to add Tanzania

Over the line: Serengeti National Park, Tanzania, at the border with the Maasai Mara National Park, Kenya, indicated here by a stone marker.

Shining light: It helps to have an asset like the 'magical Kenya' brand essence, captured in this sunset shot with a buffalo's silhouette at Amboseli National Park.

to our total lifetime list of countries and territories that we have visited is a question we are still discussing—does such a brief stop count?

The Serengeti stretches over 30,000 sq km, about twenty times the size of the Maasai Mara. 'Serengit' means 'endless plains' in the Maasai language, Maa. The two areas are joined by the Mara River, the scene of the most dramatic action of the great migration, which is what draws most tourists to this area. The Maasai Mara has a higher density of trees and a much more hilly and undulating landscape. With the contrast in physical size, the Maasai Mara might have cause to be a little bit humbler in the presence of the Serengeti, but no, the Maasai Mara more than holds its own—and for good reason.

Once again, Kenya has an edge over other African nations, with factors including its richer and more significant colonial history, its economic heft and its international connections as the largest economy on the continent. Overall, brand Kenya is stronger, reinforced by more impactful representations, for example, depictions and citations in literature, cinema and pop culture. For Tanzania, or any other challenger, it would take some effort to dislodge the primacy of recall and affinity for the brand attributes associated with all the notions around 'magical Kenya'.

Spain

The icon as catalyst

The haiku poem and ink sketch in this haiga artwork are inspired by what the Guggenheim Bilbao represents for Spain: A showcase with a daring new vision arising from a storied past, the abstract outline of a ship recalling the heritage of this port, but with the sharp design of a vessel for the future, to set off on many journeys of the imagination to faraway places like China, because no physical distance could, or should, cut off the shared sense of humanity.

SPAIN

Global scope

this space vessel sails,
shines, even with journeys launched
from most distant tales

許木桧
2020

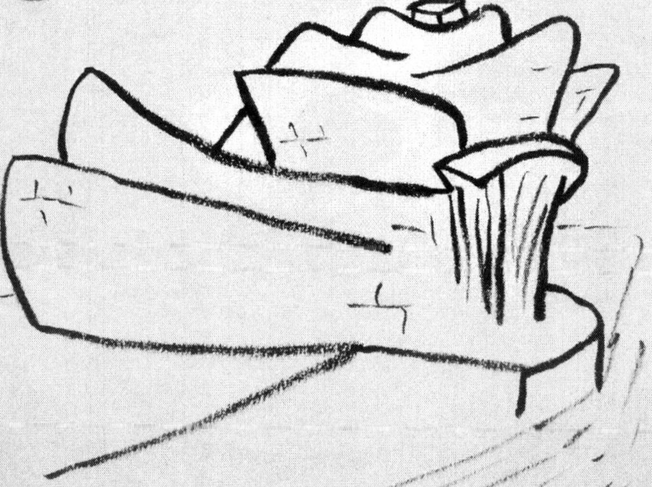

Brand power: The Guggenheim Museum revitalized Bilbao, and with it, also the transformative potential of the cultural icons of other cities.

The 'Bilbao effect'

From Africa, we flew into the cross-straits of Europe and Asia. From Nairobi, Kenya, into Barcelona, Spain, we transited through Istanbul, Turkey. We had both visited Istanbul once before in the 1990s, my wife on holiday, and I on a journalism assignment to Turkey and Egypt. For me, Istanbul is a very special place as the confluence of Europe and Asia beside the waters of the Bosphorus River. Its unique interface between East and West made this transit stop, in some ways, the perfect amuse-bouche for the next course of the cultural salad mixing bowl that Spain offers.

Sometimes, when a name becomes an adjective in global discourse—as with the 'Bilbao effect'—it is clear that an idea has captured the world's imagination. The transformative power of culture in Bilbao's place-making can be seen not only in tangible data such as visitor arrivals but, more importantly, in intangible aspects such as the psychological effect of enhanced civic pride. The Guggenheim is to Bilbao what to

The 'Bilbao effect': Catapulting the city from the fourteenth into the twenty-first century, like the bright lights illuminating this night view of Bilbao City Hall.

The 'Guggenheim effect': Other cities can see reflections of their own future selves through iconic rejuvenation in the mirrored installation '80 Balls' by Anish Kapoor, one of the artworks around the Guggenheim Museum's riverside promenade, adorned with flowers in spring.

Barcelona are its stunning buildings of Modernist architect Antoni Gaudi and Barcelona Football Club, home ground of football superstar Lionel Messi, and to Madrid, its Baroque heritage such as the Museo del Prado, and Cristiano Ronaldo's Real Madrid.

In the 1920s, the Irish novelist Kate O'Brien, who lived in Bilbao for some time, said that this was a place where no real tourist ever went. And that was generally true, even as recently as the 1990s. But ever since the Guggenheim Museum opened in 1997, the city can claim top billing as one of the biggest cultural magnets in Europe, largely because of the impact of this one extraordinary building. The structure has been featured in the media countless times, including in the opening sequence of the 1999 James Bond film *The World Is Not Enough*.

'Icon' is something of an overused word, devalued by careless usage, like so many other words, to mean almost anything representative of something else, and available to be attached to any person and their soft toy. Strictly speaking, at its best, 'icon' describes something exceptional, something worthy of veneration. And truly, Bilbao's Guggenheim restores the weight of this word. In place-branding terms, Bilbao has won for itself a special position, to have its name attached to an 'effect' that other cities look up to.

Of course, Bilbao is not the first example of how a cultural icon can capture the world's attention and remake a place almost completely. The best-known such case study of the twentieth century, arguably, is the Sydney Opera House in Australia. Initially dogged by controversy and negative reception, the arts centre has come to represent not just Sydney but the whole of Australia as a country brand, with the opera house's 'orange peel' silhouette adorning many a fridge magnet or tote bag.

In place-branding terms, Bilbao has won for itself a special position, to have its name attached to an 'effect' that other cities look up to.

If one were to go further back in history, possibly there could be other examples, such as the pyramids of Giza, Egypt. But, with the dramatic 'before-and-after' story of the Guggenheim here, Bilbao possibly seized international attention more quickly and impactfully. If Sydney has the most famous transformative arts centre case study, then Bilbao has that mantle for museums.

The Guggenheim is, without doubt, a stunner. It was difficult to peel our eyes away from it, from the first fleeting glimpse of its steely exterior from our passing taxi from the airport late at night, to the last gaze upon it on our last evening in Bilbao, as the light turned from dusk into twilight. This architectural marvel on the bank of the Nervión River, designed by the Canadian-born American architect Frank Gehry, has an effect like a catalyst in chemistry, a substance sparking other chemical reactions that have their own impetus and momentum. The opening of the Guggenheim Museum in Bilbao turned what was a struggling, mostly forgotten, industrial town into a thriving cultural metropolis that draws visitors from all over the world and inspires countless city planners, cultural practitioners and aficionados to dream big.

Such high regard became even more widely held when Bilbao was recognized in 2010 with the Lee Kuan Yew World City Prize, awarded by Singapore, for being a model of urban development, for following through on the 'Guggenheim effect' with good city planning, governance, implementation and maintenance. This encompassed many facets of sustainable development across the city, including enhancing broad riverside promenades and promoting a culture of cycling and walking over driving. For me, Bilbao was well-known from my work on liveability with the Centre for Liveable Cities in Singapore, so Bilbao was high on the priority destination list for this journey. We had to include the European continent on our worldwide itinerary, so we had to pick at least one stop. Because we had travelled extensively in Europe previously, our focus for this trip was on other regions of the world. Bilbao was, for me, a must-visit city for its place-branding significance, so Spain became the only European country on the itinerary.

Bilbao is a huge success for Spain's nation-branding project Marca España (brand Spain), despite all the attached political complications. Bilbao is the capital of the autonomous community of the Basque Country. Bilbao's spiritual sibling is Barcelona, capital of the other main autonomous community of Catalonia, which has, on many occasions, asserted its independence from the central authority of Spain. The brand essence that Bilbao has built up as an innovator for liveability is an example of how brand Spain could do much more in terms of projecting its soft power. This adds a vital dimension to what the country already has in its rich reservoir of cultural resources, which includes artists such as Pablo Picasso, founder of the Cubist movement, and writers such as Miguel de Cervantes, known as the greatest author in the Spanish language for novels including the classic *Don Quixote.*

With the city's geographical proximity to Barcelona—a forty-two-minute flight away—it is no coincidence that Bilbao also shows the effect of what has been called the 'Barcelona model'. This model—of

Global intersection: On a stopover at Atatürk Airport in Istanbul, Turkey, en route from Nairobi to Barcelona, figurines in an airport shop of Barcelona FC soccer stars Lionel Messi, Luis Suarez and Neymar attest to the global soft power of brand Spain as a launch pad for global talent.

an entrepreneurial city that can significantly manage cultural resources for the benefit of city branding—can trace its inflection point back to Barcelona's hosting of the 1992 Summer Olympics. The Barcelona Olympics' sporting showcase catapulted to prominence all the attractive qualities of a modern Spain, refreshing and revitalizing brand attributes such as 'flamboyant passion', whether expressed in the feminine flamenco dancer or the macho bullfighter. Bilbao rode on the legacy momentum from Barcelona 1992 to remake itself and earn an enduring position amongst the world's top place brands.

Here in Bilbao, one finds the impact of disruptive innovation on a grand scale, that has an intriguing history. Bilbao, located in the north of Spain on the Bay of Biscay, had been the most important industrial seaport in the Basque Country from the nineteenth century. With its strategic access to the Atlantic Ocean, Bilbao's economy—based

Puppy love: This thirteen-metre-high topiary sculpture of a West Highland White Terrier by the American artist Jeff Koons, thought of as the Guggenheim's 'guard dog', was the site of a foiled assassination attempt on the Spanish king at the museum's official opening in 1997.

on shipbuilding, and coal and steel production—was booming until the 1970s, when dockyards and factories had to close down, faced with changes in technology and supply chains and competition from other ports.

The doldrums dragged on until the beginning of the 1990s, when the American Guggenheim Foundation, in search of a location for a European museum, was in negotiations with several important cities. Bilbao's authorities showed the strongest interest in the project, seeing the prospects of a renowned 'temple of art' for a city in decline. Perhaps being a relatively smaller town, Bilbao inherited less of the focus on the self that comes naturally to citizens of imperial power. The eventual deal was simple: the Basques would provide space and funds, while the Guggenheim Foundation would fill the museum with works from its prestigious collection and handle its management.

Meeting the opposition to a 'McGuggenheim'

There was no lack of opposition, initially. Many residents of Bilbao protested against the project, baffled at why their city would invest millions in a museum—a luxury item—instead of on the more urgent necessity of reviving factories to help the local population. Bilbao's shipping industry workers, like coal miners elsewhere, had been displaced by automation and were demanding rescue by the government. With this shift from commerce to culture, many families must have relocated to rebuild their lives elsewhere. But many others have since moved in, as cultural tourism brought a boom to many other industry sectors, from retail to restaurants to real estate. The larger backdrop of Bilbao's transformation is the great shift from the dominant economic activity of manufacturing towards services, on the road towards the current Fourth Industrial Revolution. At a time of Covid-19, Bilbao will still hold its appeal for domestic tourism, until it can bounce back with an eventual recovery.

Also, as a reflection of the independent spirit of the Basque Country, the Guggenheim arrangement was criticized for being a 'McGuggenheim', an act of commercialized cultural imperialism paid for by the people. Days before the opening in October 1997, a group of Basque separatists posing as gardeners planted a dozen remote-controlled grenades in flowerpots at the paws of Puppy, a thirteen-metre-high topiary sculpture of a West Highland White Terrier by the American artist Jeff Koons. The intention was to kill King Juan Carlos I and Queen Sofia of Spain during the opening ceremony. Thankfully, the plot was foiled, although a policeman was shot dead, and this incident has even been mostly forgotten. Ironically, Puppy—star of millions of Instagram photos—is now thought of as the Guggenheim's 'guard dog'. The day we visited, we spent almost as much time admiring the exterior of the Guggenheim as we did exploring the interior. The time outside included enjoying a sushi lunch at a restaurant perfectly located directly opposite Puppy, with a full view of the museum's entrance plaza.

What the success in Bilbao shows is that a place-branding hit is always a package deal—it is the totality of Bilbao that makes this Guggenheim what it is, more than just a Guggenheim on its own merit.

The opposition to the Guggenheim had a point, but here is yet another struggle between purity and pragmatism, and there are no prizes for guessing which one won in the end. In order to garner international appeal, there is really no choice—given the global dominance of the Western art world—but to link up with brand names. This approach is also being adopted by places such as Abu Dhabi, which already has its successful Louvre Abu Dhabi museum, even as the Guggenheim Abu Dhabi project is said to be moving forward again from 2019, after it was stalled following its first announcement in 2007.

Amongst the architectural community, the design innovations of Bilbao's Guggenheim museum were showered with praise. 'It is the greatest building of our time,' said the US architect Philip Johnson, adding: 'When a building is as good as that one, f**k the art'—this is an exaggeration, no doubt, but it is nonetheless a statement that captures the power of external form to provoke a response. In one interview, Gehry emphasized an aspect of the design often overlooked by critics and imitators alike, which is that it is not purely fanciful but rooted in the reality of its context, working hard to connect with its surroundings: 'I spent a lot of time making the building relate to the 19th century street module and then it was on the river, with the history of the river, the sea, the boats coming up the channel. It was a boat.' The building itself is worth a visit, even without closer inspection of the art, and this is a key aspect, which is also true of other city/country symbols like the Eiffel Tower.

However, it is only when form and substance coalesce in a structure that it can gain admission into the premier league of icons. The Bilbao Guggenheim shows off Gehry's signature style, and this aesthetic approach can be seen in his other works, such as the Biomuseum, a biodiversity museum in Panama City, Panama. But Panama's Biomuseum has earned only a fraction of the mindshare and drawing power of its elder sister structure in Bilbao. What the success in Bilbao shows is that a place-branding hit is always a package deal—it is the totality of Bilbao that makes this Guggenheim what it is, more than just a Guggenheim on its own merit.

> The Guggenheim Bilbao can play a vital role in deepening cross-cultural understanding—a crucial endeavour that exemplifies art's greatest contribution to humanity, which it has been called upon to justify and reaffirm in this time of Covid-19.

There are, of course, minus points to the Guggenheim museum, if one looks hard enough. Chief amongst these is that the pedestrian access to the building could certainly be improved. There is inadequate signage—in the time that we were there, we saw a few tourists making U-turns because they had gone the wrong way while trying to locate the main entrance. And even when one is headed in the right direction, the steep, irregular steps leading to the museum entrance are not the most user-friendly.

However, such shortcomings aside, one can only be impressed by the eye-catching, flowing curves and iridescent sheen of the museum's titanium exterior. Ultimately, however, such a project can truly work only if the contents also resonate well with audiences. And this is where the most fascinating questions centre on Bilbao's Guggenheim.

The world comes to Bilbao

'To learn is to know one's ignorance,' Confucius said. The Guggenheim's positioning amongst the elite in international arts circles is evident from its outdoor sculptures, with works by artists from Jeff Koons to Anish Kapoor. But the true value of an arts institution is the extent to which it can go deeper to serve as a place of learning, living out its declared openness to ideas from elsewhere around the globe. The exhibition that we saw demonstrated this abundantly. Quoting an ancient Chinese sage at the start of this section seems all the more appropriate, as, at the time of our visit, the blockbuster show *Art And China After 1989: Theater Of The World* had just moved to Bilbao from its first showing in the Solomon R. Guggenheim Museum in New York City.

This major showing of art from China to audiences from Spain and the rest of the world visiting the Guggenheim would be, for some, their first major introduction to modern Chinese art and to some of the social and political issues touched on. The morning we visited, we seemed to be one of the very few, if not the only, ethnically Chinese visitors there at the time. Naturally, some of our Caucasian fellow-visitors might have assumed we were Chinese nationals. A few might have been tempted to ask questions prompted by what they had seen in the exhibits. Once or twice, I was a bit distracted, wondering what was on their minds.

The Guggenheim Bilbao, it seems, can play a vital role in deepening cross-cultural understanding. This is a crucial endeavour that exemplifies art's greatest contribution to humanity, which it has been called upon to justify and reaffirm in this time of Covid-19. Art's value-add is like tossing something, as it were, into that gulf of different perspectives that one can reach across only with empathy, if only this capacity existed in more people. This exhibition provided a wide array of insights into the cultural transformations at work in China during a crucial phase of history, with consequences reaching into the present day. The period between 1989 and 2008 book-ended two decades of intense change in a country that is set to be a dominant force in the world for a long time to come, even with the disruptions brought about by Covid-19.

As soon as we set foot in Spain, we had already practically been greeted by this exhibition. The show was being promoted right next to the Barcelona airport's luggage carousel, with ads showing dissident Chinese artist Ai Weiwei's *Dropping A Han Dynasty Urn* (1995). The

Chinese takeaway: This 2018 exhibition on China leaves impressions that frame foreign perceptions of the Middle Kingdom, through works such as Ai Weiwei's *Dropping A Han Dynasty Urn* at Barcelona airport and Wang Guangyi's *Mao Zedong: Red Grid No. 2* reproduced on a poster by the Nervion River.

work—showing three photos of Ai dropping an urn—can be seen as a kind of triptych statement of loss and regret from the social changes underway in China. Ai, China's best-known artist-political activist, lived in the US from 1981 to 1993, before returning to China, then moving to Germany and on to Britain. In China, he helped establish an experimental artists' community at Beijing East Village, and artworks from this group figured prominently in this Guggenheim exhibition.

Maritime influence: Bilbao's Las Arenas Coast, with its murals of the place's shipping and fishing past, complete with heritage write-ups along the promenade and views of the seaside mansions of the old Spanish aristocracy.

Three of the works that had caused a huge controversy in New York were shown in Bilbao, to less adverse reaction. Huang Yong Ping's *Theater Of The World* (1993), the work from which the exhibition borrows its title, has a cage in which live insects, snakes and lizards fight for survival. This work can be read as a depiction of the viciousness of the world and a commentary on the state of society in modern China, and of humanity in general.

Xu Bing's *A Case Study Of Transference* (1994), which was censored in New York City, was shown in full here, but only on film. In the original performance, the artist placed two live breeding pigs—a male boar stamped in ink with gibberish in Roman characters, and a sow stamped with illegible, invented Chinese characters—intended as a satirical cultural allegory of what happens when East and West come together.

Art shows like this open up an incredibly valuable window into foreign cultures, which most visitors would normally only encounter through perceptions carried in the mainstream media and then amplified on social media. With mass travel curbed by Covid-19, displays of art like this China exhibition become even more crucial for cross-cultural understanding. This is the greatest value of an iconic structure—it draws people from all over the globe together to get a glimpse of other worlds and cultures, and hopefully in the process, see what is common and shared.

Aside from the area around which the Guggenheim Museum is sited, Bilbao has other neighbourhoods that are well worth visiting. One of them is the Las Arenas district on the Spanish coast. Part of the journey there by public transport includes going on the Vizcaya Bridge, which was declared a UNESCO World Heritage Site in 2006 for being one of the most outstanding buildings made of iron during the Industrial Revolution. It was opened to the public in 1893 as the world's first mechanical bridge, and is still working today. It is like a skytrain in an airport terminal built over a river, except that the suspended 'capsule' of this bridge only moves to and fro across the mouth of the Nervion River.

Here too, one can take a very pleasant walk along the promenade and take time to admire the old seaside mansions of the landed gentry of the early twentieth century. Some of these homes had been lived in by the same families ever since they were first built. Some were inspired by sixteenth century palaces; others also housed items of historical interest such as private libraries. These are relics of the inequalities of the past, and they make for absorbing observations of sociological interest. For dinner, we picked an ancient tavern with a view of the coast that offered a satisfying end to the evening's stroll. Luckily, they served pub food so we could eat before sunset, as most Spanish taverns serve full meals only much later in the night. A delicious Spanish tapas dinner for the body complemented well the satisfying meal for the soul that we had enjoyed earlier that day in absorbing the cultural power of Bilbao.

A superb side dish: San Sebastian

Like a flavourful bite-size side dish that enhances the main course, visits to major cities are all the more fulfilling when they also offer a delightful side trip. Spain has wonderful options, like a generous tapas spread of glorious places to visit. Barcelona's attractions include, in its hinterland, Montserrat, a rocky mountain range in Catalonia, whose jagged ridgelines are, I believe, one of the artistic inspirations for the Catalan architect Antoni Gaudí's much-loved asymmetrical, undulating roof designs in Barcelona, including his most famous work, the church of the Sagrada Família. Madrid has Toledo, known for the medieval Arab, Jewish and Christian monuments in its walled old city. For Bilbao, the best side trip is to San Sebastian.

About 100 km from Bilbao, San Sebastian (known locally by its old name Donostia) is an easy bus ride away of just over forty minutes. The views from the surrounding ridgelines are spectacular. From there, once you have put your bags down at your hotel or other lodging, La Concha Beach is extremely inviting, just to stroll along the promenade, have a drink or snack at a quaint café, or walk all the way through to the old town, and on up to La Mota Castle, where you can go right up to a Christ statue that you would have seen from a distance earlier from almost everywhere else around.

San Sebastian is close to the French border and has its own distinctive language and culinary traditions. It is set apart physically from France,

Basque delights: For cuisine, a gourmet's paradise of Michelin-star dining, and for scenery, gorgeous sunsets like this one over the promontory of Monte Igueldo.

for sure, and also culturally from the rest of Spain in some ways. A dining experience not to be missed is to have a meal of pintxos, which are almost unique to San Sebastian. They are called pintxos (in Basque spelling) because many of them have a pincho (Spanish for 'spike' or 'pierce'), typically just a toothpick through the food items. Their smaller size and particular varieties are what make them slightly different from the more widely known tapas found elsewhere in Spain.

The Basque Country is a gourmet's paradise, being home to the world's second-most Michelin-star restaurants per capita, pipped only by Kyoto, Japan. In 2018, pintxos were ranked top in *Lonely Planet*'s list of the 500 best food experiences in the world. This gastronomical bucket list of meals features choices from some of the world's best chefs, and, for foodies, is worth building entire vacations around.

In Spain, dinner typically starts at 9 p.m. or later. One of the plus points about pintxos is that they are more like appetizers to be enjoyed with pre-dinner drinks, and so they are available earlier at what would

Bite-size culture: Pintxos—a unique dining experience in San Sebastian's old town, Parte Vieja (Old Part), against a backdrop of ancient architecture.

Above the conch: The Christ statue at La Mota Castle on Monte Urgell seen from afar at San Sebastian, and closer from La Concha Beach.

be considered a 'normal' dinner time by most people from elsewhere in the world. That night, we were famished by the time we picked a bar to settle down at. The best area in San Sebastian to try pintxos is in the old town, Parte Vieja (Old Part), famous for having the highest concentration of bars in the world.

We settled at one of the favourite pintxo bars there. Just surveying the range of items available can make one heady, even before a single sip of txakoli, the slightly sparkling, very dry white wine that is a favoured liquid accompaniment for pintxos. The Basque essentials on offer included succulent seafood such as gratinated crabmeat, octopus, prawn and sardines, paired with wild mushrooms, eggs and vegetables from the family farm. The bar we were at is located just down the street from the Santa Maria Church, whose steps provide overflow seating for the bars around. It is like a carnival of snacks, spread over the whole old town.

Pintxos have a bit more distance to close to become a true culinary icon, but they have a solid foundation and launch pad in San Sebastian and the Basque country.

In recent years, pintxos have started to travel around the globe, reaching as far as Singapore, where the first pintxo bar opened in 2019 at a beautiful venue—the heritage building Alkaff Mansion in Telok Blangah. When we visited this bar, we could not help indulging in a little stretch of the imagination, as this building did remind us of the mansions of Las Arenas in Bilbao. This made this location all the more impressive in bringing a bit of San Sebastian out so far from its home.

Pintxos have a bit more distance to close to become a true culinary icon, but they have a solid foundation and launch pad in San Sebastian and the Basque country. For now, domestic tourism will be more than enough to sustain all the customs surrounding pintxos, until, after the worst of Covid-19, they can resume flourishing internationally. Everything about pintxos—not only the food itself and the accompanying drinks, but also their history and cultural connotations and significance—makes up yet another brand offering to add to the already formidable power of brand Spain. There may never be another Guggenheim in Spain, but we will certainly witness the Bilbao effect some more for a long time to come.

Colombia

Culture as power

The haiku poem and ink sketch in this haiga artwork highlight a favoured style of the art of Fernando Botero: His distorted, larger-than-life sculptures and paintings, some of which can be read as critiques of various human excesses, while others are emblems of issues that he chose to champion. His cultural contributions are a key facet of the place brand of Bogotá and Colombia, a gift and inspiration to all mankind.

COLOMBIA

Larger concerns

engorgements spotlight
other symbols, inflations:
causes for the fight

許木松
2020

Riding high: Colombian artist Fernando Botero's works reaffirm how a culture can ride art's global appeal to shape a place and a people in the world's imagination.

The oversized power of culture

Culture is what makes a place and a people. It is the substance, the core, the essence, of soft power. On this score, one of the most distinctive and fascinating case studies is Colombia. This South American nation's cultural superhero is none other than the artist Fernando Botero. He represents not only his country but also an entire subcontinent and civilization, as one of the most recognizable faces of South America.

Inevitably, and totally unapologetically for us, Botero occupied centre stage on our very limited agenda in our all-too-brief stopover in the Colombian capital of Bogotá, en route from Barcelona, Spain, to San Jose, Costa Rica. Wanting to get a feel of the soul of Colombia through his art and artistic vision was a priority—to try to get a glimpse and a deeper sense of the context in which he lived and created most of his works. This is the value of the visual arts, to open windows into the soul of a whole society. This is why visiting an art museum is a key highlight of a visit to any city, and why engaging with a worthy artwork is always an ichigo ichie encounter.

Culture is what makes a place and a people. It is the substance, the core, the essence, of soft power. On this score, one of the most distinctive and fascinating case studies is Colombia.

As Botero himself once said: 'A painted landscape is always more beautiful than a real one, because there's more there. Everything is more sensual, and one takes refuge in its beauty.' On this day, in front of us, the landscape that Botero spoke of was made of the colourful colonial houses and narrow cobbled walkways of Bogotá, part of the spiritual resources that must have informed all of his art, at the deepest levels of his subconscious. Our time in Bogotá was limited, but we knew that it would be a precious experience just to get a feel, however fleeting, of this land that has produced such impactful culture.

We felt steeped in culture as soon as we checked into our accommodation for the night and looked out of the room window to see the terracotta roofs of the surrounding buildings sloping up to the nearby Cerros Orientales (Eastern Hills). Our hotel was in the same

Botero writ large: Clockwise from top left—Botero by the Singapore River; at the Singapore Art Museum as part of an extensive exhibition in 2004; in the Praca do Comercio in Lisbon in 1997; and at the China Art Museum in Shanghai in 2016.

precinct as the country's national theatre, the Teatro de Cristóbal Colón (Christopher Columbus Theatre), known as the 'Teatro Colón' for short. Botero filled our imaginations; his sculptures no doubt danced through our dreams that night.

To say that Botero's reputation precedes him is an understatement as large as his oversized sculptures. For us, Botero's works had already left an indelible impression previously on our travels, when we had admired his works in places such as the main city square of Lisbon, Portugal; in the galleries of the Museum of Fine Arts in Boston, Massachusetts, USA; and the entrance plaza of the China Art Museum in Shanghai, China. In Singapore in 2004, seeing his works in several prominent outdoor places, including at the old Esplanade waterfront and in front of the Singapore Art Museum, was an artistic treat. The Singapore show remains the most extensive sculpture exhibition ever of a single artist's work in the republic's history.

In November 2004, I had the chance to see Botero himself in the flesh during his visit to Singapore. I was attending a public dialogue with Botero alongside the Taiwanese sculptor Ju Ming at the Suntec City convention centre, as part of his Singapore show. Botero was asked by a member of the audience whether his art was a critique of his country. His reply: 'Social commentary belongs in the newspapers'.

As a newspaper journalist then, I remember sensing that he seemed to be downplaying the commentary in his work, for reasons best known to himself. Perhaps, as a guest in a foreign country, he preferred—like many visiting artists do—to avoid comments that might be construed to be political. Absence of social commentary might be true of most of his sculptures of figures such as birds and cats, but this point certainly does not apply to his entire oeuvre, unlike for some other visual artists.

In fact, a year earlier in 2003, Botero had been exhibiting and speaking about a set of works he had been preparing since 1999. For the first time, his works openly took on the drug-fuelled guerrilla war that had been going on for forty years in Colombia. The paintings and sketches, with titles such as *Death In The Cathedral*, show mass funerals, bombings, massacres and suffering caused by rebels and paramilitary forces, with all the artworks depicting only victims but not killers.

Spiritual home: Museo Botero, a must-visit place in Bogotá.

At that time in 2004, Botero would have already started on an even more overtly critical series—exhibited from 2007 onwards—of 'Abu Ghraib' drawings and paintings about the torture of Iraqi prisoners by American security personnel at the Abu Ghraib detention centre, west of Baghdad, Iraq. As Botero said at a dialogue at the University of California Berkeley: 'The power of art is to make you remember something.'

Despite his own disavowal of it at that event in Singapore, commentary was indeed richer in his paintings on display at the Singapore Art Museum, such as *The President And First Lady*. By simply representing these figures in regalia, aloof on horseback, with a backdrop of plantation banana trees, Botero highlights the divide between authority figures and followers.

After listening to his talk, I wrote a column on Botero in a series on leadership that I was contributing at the time to *The Straits Times*, the main English newspaper in Singapore. This op-ed piece, on the relationship between art and leadership, closed on this observation:

Big in Bogotá: Naturally, Botero souvenirs are big (pun intended) in Colombia, as in this tourist shop in the capital.

'There is more value in the kind of inspiration of a Botero—truthful yet non-judgmental; disturbing sometimes, but always ultimately uplifting. This, more than anything else, should be the greatest gift of art, and of leadership.'

Botero's signature style, known in the art world as 'Boterismo', depicts usually human or animal figures in exaggerated size, which are described with adjectives such as 'voluptuous' or 'chunky', depending on the piece and the viewer's mindset or mood. It is an illustrative method that works extremely well—it gives licence for works to be larger than life, hence expanding their significance by virtue of size alone, if nothing else. It also lends satirical scope to a work, say, of a political figure, as if the engorged volume were mocking a flaw such as an inflated ego.

In his homeland of Colombia, it is no surprise that Botero would have an industry sub-sector for tourist merchandise. If anything, what might give one pause is that thriving sales of all things Botero do not seem to be more prevalent and prominent. There are only a handful of shops that sell Botero-related items, when one would have expected more. As for the shops at places like the airport, these are clearly better

Old Spain: View of La Candelaria old quarter and the Monserrate mountains on a rainy, foggy morning, and the Primatial Cathedral, rebuilt three times on the spot where Spanish conquerors had erected a thatched chapel.

stocked with materials from, or inspired by, official national country branding campaigns. Perhaps this is an indication of the relatively lower volume of foreign visitors to Colombia compared with some other cities associated with major cultural figures. Covid-19 would have trimmed tourist arrivals even further, but this will just be a temporary interruption to the global spread of Botero's cultural power. In fact, it will be the global draw of Botero that will help Colombia boost its economy for the post-pandemic recovery.

Botero's adoption into the 'Western world'

In the international art world, there is no escaping the predominance of the 'Western world', meaning primarily Western Europe and North America, and this is also true here. The Museo Botero, a must-visit in Bogotá, has 123 works of Botero and eighty-five of other artists. Here, the first room is dedicated to the all-time Western greats, including Pablo Picasso and Salvador Dali. One of the works, *La Mona Lisa*, is a depiction of the famous Leonardo da Vinci painting in Botero style with the face in expanded proportions but with tiny eyes, nose and mouth—a parody, in a way, but also a compliment in recognition of one of the world's best-known works of art. Botero himself has been 'adopted' into the elite of the Western art world, surmounting the cultural gulf

Written in history: Gabriel García Márquez—always remembered at the cultural centre that bears his name in Bogotá.

between South America and the main spheres of Western art in Western Europe and North America.

This power dynamic of the global art world has to be appreciated in the wider context of Colombia being a former colony of Spain. Monuments of this colonial history are everywhere in Colombia. In Bogotá, the chief example is the Metropolitan Cathedral Basilica of the Immaculate Conception, a Roman Catholic church located on the eastern side of Bolívar Square. Its Spanish architecture dominates the whole vicinity, and it is, naturally, the spiritual heart of the city. Religion is one of the twin pillars, together with language, of the legacy of empire. We were visiting early one morning amidst a slight downpour, so it was practically deserted except for people heading to work, with more pigeons than humans about. The rain gave added poignancy to this square that has witnessed so much history. On weekends and festive occasions, though, it would be packed with people at this social gathering place for the whole city.

This power dynamic of the global art world has to be appreciated in the wider context of Colombia being a former colony of Spain.

As a cultural figure in Bogotá, alongside Botero is the other artistic giant of Colombia—the celebrated writer Gabriel García Márquez, whose immortality is assured in the global literary world. García Márquez is best known for novels such as *One Hundred Years Of Solitude* (1967) and *Love In The Time Of Cholera* (1985). His works have achieved significant critical acclaim and widespread commercial success, most notably for popularizing a literary style known as magical realism, which uses dream-like elements and events in otherwise ordinary, realistic situations. When he won the 1982 Nobel Prize in Literature, he said in an interview: 'I have the impression that in giving me the prize, they have taken into account the literature of the sub-continent and have awarded me as a way of awarding all of this literature'.

Whether Botero or García Márquez wields the greater cultural influence is arguable, given the writer's relatively more educated, and therefore smaller, audience. But just taking a simple barometer test of unpaid viral marketing might give a perspective on this question—whereas Botero's sculptures are enjoyed every day by thousands of people around the world, from Santiago to Shanghai, who walk past and take selfies with them, thus going on to share them with even more people, definitely fewer people are quoting the words of García Márquez in their social media posts.

This distinction in scale is also indicated by the different forms and levels of these two artists' representation in Bogotá. Whereas Botero's works and scores of replicas can be found in the handful of merchandise shops in the city, García Márquez's legacy is harder to see, as it is even more self-contained, mainly in and around the Cultural Centre that bears his name, near Simón Bolívar Plaza in the La Candelaria district. Thus, more promotional effort is called for from García Márquez's audiences. One of the best-loved quotations of García Márquez is this: 'What matters in life is not what happens to you but what you remember and how you remember it.' At the Gabriel García Márquez Cultural Centre, which has an art gallery and library, visitors must make the effort to form their own memories from his works to remember him and his artistic legacy.

The art of García Márquez—in its mixing of the magical and the mundane—gives deeper insight into the condition of Colombia's soul-searching and quest for identity. 'Surrealism comes from the reality of Latin America,' he told *The Atlantic* magazine in 1973. Some writers influenced by García Márquez, such as Salman Rushdie, argue that magical realism is a specifically post-colonial literary technique and that it works to its fullest effect in interaction with countries that have been colonized.

García Márquez's magical realism 'expresses a genuinely "Third World" consciousness', wrote Rushdie in a 1982 review of the novel *Chronicle Of A Death Foretold.* He explained: 'It deals with what [V. S.] Naipaul has called "half-made" societies, in which the impossibly old struggles against the appallingly new, in which public corruptions and private anguishes are more garish and extreme than they ever get in the so-called "North" [the world of the former European colonial masters], where centuries of wealth and power have formed thick layers over the surface of what's really going on.'

Spell it out: This 'official message' in a Bogotá shop window shows that Colombia's country branding has its work cut out for it, when it has to start with getting its name spelt right with target audiences.

Left to their own devices, colonized peoples would, in due course, also develop those 'thick layers' of denial of reality that Rushdie wrote about. Usually, it is only art and cultural leadership that can pierce or pull away those layers, as if to push aside a massive quilt that would smother everyone and everything beneath if left unmoved. Magical realism helps to question, and subvert, the more recent realities, hence its value in examining the legacies of imperialism.

Please, it's not Columbia

To command attention and shape perceptions, culture can play an even larger role in countries that do not have the same top-of-mind recall as those that have the world's most powerful nation brands. For Colombia, in the first place, there is a more basic problem to contend with—many people cannot even spell its name right.

The problem arises mostly with the key tourism segment of visitors and audiences from North America, some of whom might be confused by being more familiar with the spelling 'Columbia', as used in such contexts as the District of Columbia where the US capital Washington is located. The sense of humour is refreshing. But still, to have to put the message 'It's Colombia, not Columbia' on tourist merchandise indicates that the country brand is in a mode of what is called 'brand lag' (in a position of having to address basic misperceptions), and thus, requiring 'brand recovery' efforts, which sets the whole enterprise of nation brand-building back even at the starting-point.

Moving forward, with the brand ambassadors it already has on the world stage, Colombia could have less of a serious identity issue if the potential of these ambassadors were more fully developed. In addition to artists such as Botero and García Márquez, Colombia has produced world-famous cultural exports including pop star Shakira and Real Madrid footballer James Rodriguez. But then, Shakira and Rodriguez made their names from overseas launch pads such as Spain, amongst the spheres of influence of the Spanish- and English-speaking Western world and internationally, where their fans do not usually think of these celebrities' nationality in the first instance.

By contrast, Botero's sculptures have a mass appeal that transcends boundaries of language and barriers of artistic literacy. He is definitely known as 'the most Colombian of Colombian artists', as he called himself in one interview. This was not an immodest boast, but a reference to the way that he seeks to draw his material and inspiration from his homeland.

A big part of Colombia's rebranding of itself is being fostered amongst the international community of liveable cities. The World Cities Summit Mayors Forum 2019 was hosted by Medellín, the capital of Colombia's mountainous Antioquia province. Medellín is making a new name for itself after winning the Lee Kuan Yew World City Prize 2016, awarded by the Singapore government, for overcoming many challenges of urban expansion and transforming itself from being known as a crime and murder capital into a much more liveable place and a model for innovation in city governance, planning and placemaking.

With the brand ambassadors it already has on the world stage, Colombia could have less of a serious identity issue if the potential of these ambassadors were more fully developed.

The innovations being applied in Medellín are mostly very progressive, including such aspects as urban renewal, participatory politics and making the city more appealing to digital natives. These are substantial improvements in quality of life that deserve to be better known around the world, especially at a time of rapid urbanization globally, and the urgency for cities to reset their priorities in the face of Covid-19.

What Colombia wishes outsiders would take the time to discover is something every country and culture would welcome—the appreciation that there are two main dimensions to every place: the traditional and the modern. There are many threads that connect Colombia to its siblings in the commonwealth of formerly colonized peoples, which have an added layer of culture to negotiate. There is their own native, traditional heritage and how to manage the many ways to move towards a modern future, that they are still in the process of figuring out. At the same time, all this is complicated by the legacies of the colonial past, what people need to unlearn before they can learn and relearn.

This is where the cultural 'products' of a Botero, García Márquez or Shakira can do much to add to the brand value of their homeland. Larger-than-life representations, weaving fantasy into ordinary existence and arresting the world's attention—these are some of the ways forward

Old and new: A romanticized image of women in the colonial past in a painting in a hotel lobby and the everyday scene of a young woman commuter at Ciudad Universitaria train station.

for a campaign that would go far, so far that the 'It's Colombia, not Columbia' campaign can be consigned to the case studies archives of nation branding.

Comfort food and cultural leadership

Generally, East Asian tourists attract attention in South America, because of the relative rarity of people of this ethnicity on this subcontinent. One of my travel memories related to this was on a boat on the Amazon River, when a Brazilian couple sitting behind us were more interested in speculating where we were from than in admiring the beauty of the rainforest around us. When we told them, we learnt that they had hardly heard of Singapore before.

I was reminded of this when we spotted an elderly couple looking at us and smiling with friendly interest when we were having a morning snack at a local restaurant in Bogotá. We smiled back politely and said hello. Because they did not speak English, and we knew no Spanish, conversation was a non-starter. But a foundation of common humanity

Colombian comfort: *Ajiaco*, a typical comfort food made the traditional way, with locally sourced ingredients.

had been established by our interest in ajiaco, a traditional comfort food of Colombia. This dish is most popular in Bogotá, possibly because the surrounding terrain was conducive for the farm production of its main ingredients, including potato. One other restaurant we had looked into along the same street did not have this item on its menu, while another only served it at lunch and dinner.

Traditionally, this dish is seasoned with a Colombian herb called guascas and is served with slices of avocado, capers and a mild sour cream sprinkled across the top of the broth. It would be a heart-warming treat (almost literally) when served hot on a winter's night. The ajiaco we had was a bit on the bland side. But its cultural significance was certainly intriguing.

The origins of ajiaco are debated, and variations of this dish have been claimed by other cultures and appear in other cuisines, notably in Cuba, where the Cubans make a heartier stew version that includes beef, pork, chicken and vegetables. The Cuban ethnologist Fernando Ortiz stated that ajiaco was a meal typical of the Taíno indigenous peoples of the Caribbean, and that it is an appropriate metaphor for Cuba being a cultural melting pot.

For a foreign cuisine to be widely embraced internationally, it would help if its parent culture already had a foundation forged through cultural leadership in broader terms.

The Colombian ajiaco is also like similar dishes in other cultures, for example the oden of Japan (a stew with radish and other vegetables) or the Welsh stew cawl, an omnivorous mix of meat and vegetables. Somehow, in Japanese cuisine, oden does not have the appeal and cache of other dishes such as sushi, while in the case of cawl, most people would not have heard of it because of the relatively low level of brand recall for anything Welsh, compared with similar items from the rest of Britain, from England or even Scotland and Ireland.

When it comes to cultural leadership through food, the dynamics are similar and yet different from, say, that of an art form like the visual arts of a Botero. Something like ajiaco would have to overcome possibly more hurdles (especially of taste and access) than a work of sculpture to claim international recognition. One of the main reasons for this is nestled in that most comforting of terms related to cuisine: 'comfort food'. A comfort food is simply something that people grew up eating. It's what your mum cooked either often, or at memorable times, like for a festival or to celebrate a special occasion like a birthday. What then usually follows is that food critics tend to deem a traditional dish 'authentic' by drawing on their own childhood memories of a particular recipe.

This is why something foreign would not gain mass appeal immediately. Some foods make it big because they were adopted by other cultures on a large scale, like Italian pizza. Otherwise, to make a foreign food become part of the mainstream menu, like Japanese sushi, it would take much entrepreneurship, investment, dedication to craft, marketing and brand-building. For a foreign cuisine to be widely embraced internationally, it would help if its parent culture already had a foundation forged through cultural leadership in broader terms. On this score, Colombia would have too much basic ground to cover first, before its food could be promoted. What it has are its cultural giants in other spheres such as the art of Botero. So, for now, for Colombia, the way to win over the outside world is not through the stomach, but the mind and heart.

Costa Rica

Small and smart

The haiku poem and ink sketch in this haiga artwork enlist the help of the strawberry poison dart frog at Arenal Volcano as an extended metaphor for a key facet of the country brand: Caring stewardship of the wilderness that has preserved this sanctuary of conservation and ecotourism in Costa Rica's forests. These frogs are lethal to predators, but bring delight if left undisturbed—just another way for a happy, enduring coexistence between man and nature.

Costa Rica

Wild essence

arrow poison no
fear — just shoot the forest dance
from afar, and go

許木松
2020

Amphibian ambassadors: Costa Rica's awesome creatures of the wild, like this red-eyed tree frog on the Osa Peninsula, symbolize the fragile value of this unique natural paradise.

Tiny and stunning

In Costa Rica, small is always beautiful. This Central American nation is home to more than 150 known species of frogs and toads of all families and appearances. Better yet—researchers believe there are countless more still to be discovered. The chances of a sighting might usually be reasonable, but there is still an element of good fortune to spot one's favourite Costa Rican amphibian and to be able to photograph it before it disappears into the undergrowth, instinctively seeking the safety of camouflage and concealment. What makes them a challenge to spot is their size, or lack thereof—some are tinier than your little finger. Some grow to at most just over 20 mm, or under an inch, in standard length. Every sighting is an intense ichigo ichie moment: 'My heart aches', as the eighteenth century English poet John Keats wrote in his *Ode To A Nightingale*.

However, any physical stature these denizens of the forest may lack is more than made up for by their stunning looks. The strawberry

Smart casual: The strikingly attractive strawberry poison dart frog, seen here in Arenal Volcano National Park, is nicknamed the 'blue jeans frog'.

Pure bliss: The customary Costa Rican greeting of pura vida ('pure life') has become an emblem featuring the sloths' carefree life of mostly slumber.

poison dart frog got its name from two aspects of its appearance. One is 'strawberry', because its torso resembles the fruit of its colour and roughly, also its shape. The other facet is 'poison dart', because, in ancient times, the indigenous peoples of the region used its powerful poison for centuries to tip their blowpipe darts for hunting. This frog is native to the tropical rainforests of Costa Rica, and also of its immediate neighbours Nicaragua to the north and Panama to the south. It forages on formicine ants and mites, which cause it to become poisonous to predators due to the alkaloid toxins produced from the chemical combinations.

The frog's toxins, if ingested by other animals, result in a disruption in cardiac function, paralysis, convulsions and death. Interestingly, poison dart frogs raised in captivity and isolated from insects in their native habitat never develop poison—an example of how adapting to life-threatening danger and other perils of certain

Blue velvet: A blue morpho butterfly on the Osa Peninsula, with wings that stay dry by making rainwater form droplets before running off the surface. This one looks like a juvenile, as its iridescent blue markings are not yet fully formed.

A simpler life: Pastoral scenes of livestock on the foothills of Arenal Volcano and cowboys riding across the plains of the Osa Peninsula.

habitats can make organisms change their appearance in strange ways. The bright and rich colours of their skin are deliberately ostentatious, a feature called aposematic (or warning) colouration, typically used by noxious organisms to ward off potential predators. This approach of conspicuity, to be seen so as to be feared, is the opposite of camouflage and avoiding being seen, the more common way of self-protection from predators.

As a reflection of more modern perspectives of anthropomorphism, this frog has been nicknamed the 'blue jeans frog' and even 'the one with jeggings'. The nickname, which arose because of its blue hind legs, never fails to bring a smile to a visitor learning of this for the first time. But such fashion monikers, while certainly cute, require suspension of the observation that this frog also wears denim full sleeves and gloves. To me, a Spider-Man costume would be a closer resemblance.

A country would have more friends and allies in times of trouble if it had already gained more conspicuity through building a recognizable country brand, making itself more obvious and outstanding, like the strawberry poison dart frog.

In some ways, what we have here is a candidate for a national metaphor for Costa Rica. This frog's method of protection against predators could work too for nations that may be threatened by powerful external forces, in a manner of speaking. A low-profile small state would find it harder to call for help should it be threatened by invasion or unwarranted attempts at domination by foreign powers. By contrast, a country would have more friends and allies in times of trouble if it had already gained more conspicuity through building a recognizable country brand, making itself more obvious and outstanding, like the strawberry poison dart frog.

What makes the condition of Costa Rica's amphibians more poignant is their vulnerability, a condition shared by all small countries. Whereas about a tenth of birds and a third of mammals in Costa Rica are at high risk of extinction, more than 40 per cent of amphibians are in similar straits—making them worse off than any other vertebrate group, according to the International Union for Conservation of Nature.

In contrast to the poisonous frogs, Costa Rica has many other wildlife species that look stunning but are not toxic. For example, the largest butterfly in Costa Rica, the Morpho Menelaus, or blue morpho, is unique because of its iridescent turquoise, almost neon, markings and large wingspan. It took us a few days of forest walks to see one, and the sight was mesmerizing. If only it had lasted more than just a few minutes.

The wings of the blue morpho have even attracted scientific research interest because of their self-cleaning properties. They have a superhydrophobic surface, meaning that when water falls on them, it forms spherical balls and rolls off the surface, so that the wings stay dry. Any dirt or dust on the surface will be removed along with the water droplets. It is one of the few butterflies that will fly in the rain. The blue morpho is also a symbol of national identity and pride: Miss Universe Costa Rica 2018, Natalia Carvajal, looked stunning in her national costume inspired by the morpho butterfly in the traditional fantasy costume parade.

A philosophy of pura vida

Ecological preservation has been a key aspect of Costa Rica's country branding, currently based on the key idea of 'essential Costa Rica'. Amongst other facets, the brand highlights the natural, unspoilt beauty of the land, truly close to the 'essence' of nature. This facet is intertwined with how the notion of 'essential Costa Rica' also refers to the character of the people. This is best summed up in the common expression pura vida ('pure life', literally, in Spanish). When asked 'how are you?', Costa Ricans say pura vida in reply. Pura vida can also be a question, to enquire: 'Is everything going well?'

This relaxed, contented life philosophy appears to be a feature, to some degree, of all tropical coastal settlements from Hawaii to Tahiti, denoting a simpler life that is generally positive, with comparatively lower levels of stress. In Costa Rica, this attitude is given distinctive character, captured in the official country branding, and of course, also used in much tourist merchandise and marketing. Apt brand ambassadors are provided by its native wildlife, in particular the sleepy

Birds of many feathers: Clockwise from top—A rufous-tailed jacamar at Arenal with a butterfly snack in its mouth, a resplendent quetzal at Monteverde and a scarlet macaw on the Osa Peninsula.

sloths, which are so relaxed that they are hard to spot, and even when located, seldom seem to lift their heads to allow for a clear enough photo that shows their faces. By contrast, the abundant birdlife of the montane cloud forest at Monteverde is unmistakable, with their vibrant hues and melodious songs.

In 2017, *National Geographic* magazine named Costa Rica the happiest country in the world, and the World Happiness Report by the United Nations placed Costa Rica as the happiest country in Latin America and twelfth on the global Happy Planet Index. Plus points include the high level of social services, the caring nature of its inhabitants, long life expectancy and relatively low corruption. As most of these rankings are self-reported states of happiness by inhabitants, they do give a clear sense of relative levels of self-perception, which is, after all, the basis for happiness. Contentment in life is probably also a foundation for honesty. Arriving from San Jose into Monteverde, I was still groggy from the flight the night before, and unfamiliar with the local currency, mistakenly gave the equivalent of a US$200 tip to our transfer driver as we proceeded to check-in. He drove back to the lodge to inform me of my error.

Costa Rica is not a cocoon, so much as a conch . . . like a seashell, waves crash all around and the tides sweep along relentlessly. But if you place your ear to the conch, you can hear the songs of the ocean, just like there is so much wonderful nature to discover and enjoy here.

On a deep level, a core trait of the people of Costa Rica is a strong sense of identity. Driving into San Jose on arrival from Bogotá at the Juan Santamaría International Airport, we asked our driver if he had heard of the 1968 song written and composed by Burt Bacharach for the American singer Dionne Warwick, *Do You Know The Way To San José*? He said 'no', which perhaps shows that the pop culture influence from the US is powerful but does not extend everywhere. This might be just one small clue, one aspect of a remarkable example of a country that has managed to strike its own path towards self-determination despite the dominance of external forces such as American soft power.

Four native species: Anti-clockwise from top—Squirrel monkey, white-faced capuchin, spider monkey and mantled howler.

Costa Rica is also a haven of political peace. Its neighbour to the north, Nicaragua, was rocked by anti-government protests from 2018 into 2019, in which hundreds of people were killed and injured. Costa Rica has been proud of not having any armed forces since 1948, after it abolished the military in the wake of victory in that year's civil war. This made it one of only a few sovereign nations without a standing army. The government budget previously dedicated to the military was allocated instead to providing health care services and education. This would go some way towards explaining Costa Rica's high rankings for happiness.

Costa Rica is not a cocoon, so much as a conch, for a country that has almost 1,300 km of coastline on the Atlantic and Pacific oceans. Like a seashell, the waves crash all around and the tides sweep along relentlessly. But if you place your ear to the conch, you can hear the songs of the ocean, just like there is so much wonderful nature to discover and enjoy here. In being blessed with the bounty of biodiversity, Costa Rica almost has it all. It possesses the greatest density of species in the world, including 840 species of birds that have been identified. While the country has only about 0.03 per cent of the world's landmass, it contains 5 per cent of the world's biodiversity. Around a quarter of the country's land area is nestled in protected national parks and protected areas, the largest percentage of protected areas in the world (the developing world average is about 13 per cent and the developed world average is 8 per cent).

Costa Rica is blessed with many incredible nature spots, like the Cano Negro Wildlife Refuge, where we saw many distinctive avian species such as the roseate spoonbill, the country's only pink bird. The unusual reptiles here include the basilisk, or Jesus Christ, lizard, which is so named because, when fleeing from predators, it can gather sufficient momentum to hold most of its body up, thus looking as if it were walking on water, as in the biblical miracle. Here, one can find what *National Geographic* magazine called the world's 'most biologically intense place'—the Corcovado National Park, the one park in the country where all four native monkey species can be found—the white-headed capuchin, the mantled howler, the endangered Geoffroy's spider monkey, and the Central American squirrel monkey, which is found only on the Pacific coast of Costa Rica and a small part of Panama, and was considered endangered until 2008, when its status was upgraded to vulnerable.

These monkeys are found also in other parts of Costa Rica, even in and around ecotourism accommodations, but usually one or two species at most. We spotted a few of them on our walks within the grounds of the places where we stayed in the Arenal Volcano National Park region and on the Osa Peninsula, where the longest hike we took stretched over 5 km. At Corcovado, however, they can all be seen, in addition to other distinctive species including birds of prey like the broad-winged hawk. Jaguars still roam here, but, of course, it would take much time and effort to come anywhere near a sighting.

What *National Geographic* magazine called the world's 'most biologically intense place'—the Corcovado National Park.

One of the mammalian highlights, the Baird's tapir, has a hide that is purplish-grey and more like that of a hippo, and is quite different from the tapirs of Malaysia, which are black and white. The drizzle on the morning we visited made it much harder to spot any birds, as they naturally were avoiding getting their wings wet, but the overcast skies also made the landscape and context more dramatic. Just before we were going to the beach to return to the boat, our guide managed to locate a Baird's tapir, foraging and chilling out amongst the mangroves by the coast. Braving the drizzle, I carried an umbrella in my left hand and manoeuvred my camera with my right hand to capture the wildlife shots we had come all this way for.

The sight was, as they say, worth the price of admission—which was not insignificant. Getting to Corcovado is quite a trek in itself, demanding the devotion of the good part of a day. After getting ready before the crack of dawn, there is the drive to the coast for the boat, which then skirts the coastline to get to the park on a small peninsula jutting into the Pacific Ocean. And only then is one rewarded with being in a unique sanctuary of nature, a haven of splendour in the grass, literally.

Buffeted by two oceans left and right, and yet sheltered from the world, such is the condition of Costa Rica, which is reflected in the

Denizens of Corcovado National Park: Clockwise from top—Baird's tapir, broad-winged hawk and collared peccary.

worldview of its artists, as described by poet and writer Luis Chaves. In one interview, he explained why Costa Rican literature must work harder to be published abroad: 'A country with no major conflicts (at least that's the idea people outside have about Costa Rica) is not attractive for foreign publishing houses.' In his poem 'Marine', he writes:

The good thing about the sea is
when nobody drowns.
It happened yesterday but
I'll tell it today
while they write
a name
on the sand
with a stick that came
floating among bags
and leaves and pipes . . .
What the sea touches
belongs to the sea.

In these lines one can hear the deep respect for the wilderness—'what the sea touches belongs to the sea'—that comes from living amidst such untouched nature. However, the reference to 'when nobody drowns' is when reality comes crashing ashore, a reminder that life always has its risks and dangers.

From agrarian heritage to new economy

The name 'Costa Rica', meaning 'rich coast' in Spanish, was, according to some accounts, first given by the explorer Christopher Columbus, who sailed to the eastern shores of this land during his final voyage in 1502 and reported vast quantities of gold jewellery worn by the natives. The name might also have come from the Spanish conquistador Gil González Dávila, who arrived on the west coast in 1522, encountered natives and appropriated some of their gold.

But, despite its name, for most of its history, Costa Rica's means have been rather modest. During most of the 300-year Spanish colonial period, this was the southernmost province of the Captaincy General of Guatemala, nominally part of the Viceroyalty of New Spain. Costa Rica was a poor, isolated and sparsely inhabited region within the Spanish Empire, because of three main factors: its distance from the centre of power in Guatemala; its legal prohibition under Spanish law from trade with its southern neighbour Panama, then part of the Viceroyalty of

Flavour of the old economy: Visit to Finca Rosa Blanca Inn and coffee plantation in Santa Barbara, with a view of San José.

New Granada (Colombia); and, by then, a lack of resources such as gold and silver.

In fact, Costa Rica was even described as 'the poorest and most miserable Spanish colony in all America' by a Spanish governor in 1719. Costa Rica was, by and large, unappreciated and overlooked by the Spanish Crown and left to develop on its own. In some ways, this proved to be a blessing in disguise, as relative neglect allowed Costa Rica to forge its own way forward. Costa Rica became a 'rural democracy' with no oppressed mestizo (mixed-race) or indigenous class.

Like a tree frog, Costa Rica, for much of its history, was not that easy to spot. Located on a narrow strip of what is today Central America, even the indigenous peoples of ancient times were rather low-key, relative to the tumult that swept through the rest of the continent. Even after the advent of Spanish rule from the sixteenth century, this territory was a peripheral colony of the Spanish empire until independence as part of the First Mexican Empire. This was followed by membership in the United Provinces of Central America, from which independence was formally declared in 1847.

Like a tree frog, Costa Rica, for much of its history, was not that easy to spot.

Today, Costa Rica is one of the most stable, prosperous and progressive nations across Latin America. After the Costa Rican Civil War in 1948, the government drafted a new constitution, guaranteeing universal suffrage and the dismantling of the military. Nowhere is Costa Rica more progressive than in environmental policies. It is the only country to meet all five United Nations Development Programme criteria established to measure environmental sustainability, and was named the greenest country in the world by the New Economics Foundation in 2009. Today, almost all its electricity is generated from green sources, particularly hydro, solar, geothermal and biomass. In 2015 Costa Rica pledged to become a carbon-neutral country by 2021.

A sip of economic heritage, harvested from the legacy of Costa Rica's old economy, can be sampled with a visit to an old coffee plantation. Coffee production played a key role in Costa Rica's history

Sports diplomacy: The Estadio Nacional in San Jose, a gift from China in 2011.

and continues to be important to the economy. In 2006, coffee was Costa Rica's number three export, after being the number one cash crop export for several decades. There are some 150,000 coffee fincas (plantations) across the country.

At the tail end of our visit to Costa Rica, we stayed one night at a Spanish-villa style inn set in a working coffee finca in the district of Santa Barbara. Our aim was to explore a bit of what one might call the country's combined colonial and economic heritage. This is one aspect of Costa Rica that ecotourists might overlook. This region gives a good sense of the charming countryside lifestyles of times gone by. We visited the coffee plantation and bought coffee (of course) as well as other souvenirs at the inn's shop. While having an alfresco dinner at the inn's terrace restaurant, we had the chance to look in wonder upon San José and its suburban settlements on the rolling foothills of the faraway mountains.

There is also a social aspect of this part of Costa Rica. This is the Central Valley, the inland region surrounding San Jose, dotted with small towns and villages. Around 70 per cent of the country's population lives around here, just a short drive from everything that the capital San José offers in terms of work and recreation.

Today, Costa Rica is a democracy that relies on the sunrise sectors of technology and ecotourism for its economic growth. The people of Costa Rica have been called upon to play more prominent roles in the development of the country, from those at the 'front office' of international tourism to everyday Costa Ricans who are the human interface, the 'crew', for the ground experience for visitors. Brand Costa Rica is in good hands, as we learned from my Costa Rican friend Daniel, whom we met for breakfast in San José. Daniel, who looks after the nation brand, shared with us his insider perspectives on the country's brand-building effort, which added depth to the impressive image we had gained earlier from projections of Costa Rica's award-winning nation brand online, and what we had seen and experienced on this visit. Yes, the reality matches the branding, which is what every nation brand hopes for.

If the country wanted to take it easy, it could just relax like a sloth and all would be well, like a real-life realization of pura vida. But since the nation has set its sights on economic growth to boost the quality of life for its people, it must constantly adapt to the external environment to survive and thrive.

With the fresh priority of growing the economy has also come a new element of pragmatism. The hotel where we stayed in San José overlooks the Estadio Nacional, Costa Rica's magnificent national stadium in Parque la Sabana. Two weeks after our visit, this sporting arena hosted a match between the national soccer teams of Costa Rica and Northern Ireland, which Costa Rica won 3-0, as a warm-up for the 2018 World Cup. This US$105 million stadium was built entirely by China as a gift to Costa Rica—one of a few clear signs, as in Kenya and elsewhere, that we saw on this trip of the shifting map of international economics and geopolitics.

Earlier, in 2007, Costa Rica had broken off diplomatic ties with Taiwan, switching recognition to China. In 2011, China became Costa

Rica's second-largest trading partner with the signing of a bilateral free trade agreement. This agreement was signed two weeks after the stadium's inauguration, marked by a match between the national teams of Costa Rica and China, which ended 2-2. There is also a domestic demographic element to relations between Costa Rica and China. Chinese residents in Costa Rica have become an important political and cultural force in the country. There are more than 45,000 Chinese living in Costa Rica, the second largest Chinese population in Central America after the one in Panama.

Country branding is important also for Costa Rica's plans to grow its economy. With the increasing importance of technology in the fourth industrial revolution, Costa Rica is right to respond by developing its hardware, know-how and international ties in technology-related sectors. But the X factor of its nation brand will always be what nature has blessed this land with—stunning natural ecosystems and awesome wildlife. These riches of nature are being taken care of well, with a government committed to protecting and advancing nature, and a culture of respecting the wild and living in harmony with it.

If the country wanted to take it easy, it could just relax like a sloth and all would be well, like a real-life realization of pura vida. But since the nation has set its sights on economic growth to boost the quality of life for its people, it must constantly adapt to the external environment to survive and thrive. It is true that there is country branding potential to be further developed in aspects of the country that have been less profiled previously, such as its heritage, in pockets such as the coffee plantations, and its people, such as the cowboys of the Osa Peninsula. But these facets can leverage off a nation brand that already has a firm, fertile foundation of nurturing nature in all its splendour. A small state must stay smart and supple—just like the tiny tree frogs in the forests of Costa Rica.

Cuba

Island of isolation and invasion

The haiku poem and ink sketch in this haiga artwork suggest how, throughout its tempestuous history, the island of Cuba has been coveted as a geo-strategic prize in the maritime theatre of operations around the Caribbean Sea. Despite centuries of Spanish colonization, American embargos, Soviet communist brotherhood and other interventions by foreign powers, Cuba's soul and spirit—given voice through music, literature and art—cannot be extinguished as easily as one could stub out a Cuban cigar.

CUBA

Cadence of hope

Carib history's prize,
incursions, cordons can't stub
your encore reprise

許木松
2020

Flying colours: Cuba's national bird, the Cuban trogon or tocororo, uniquely sporting the red, blue and white hues of the national flag, at Buena Vista coffee plantation, Las Terrazas.

Salsa and the spirit of Cuba

In all of us is a bit of Cuba. Just a bit, maybe, for some of us, but enough to make anyone hum along to a Cuban-inspired melody, enough to tempt even the most introverted amongst us to shed our inhibitions and take to the dance floor. Such is the power of Cuba's rhythm, its brand essence. It is always alluring, can be intoxicating, and has permeated the world's consciousness.

Cultural magnetism was certainly in force that evening on our night out in Trinidad, a UNESCO World Heritage town in central Cuba on the southern Caribbean coast. Casa de la Música (House of Music)—considered one of the hottest music spots in all of Cuba—is located amidst a long flight of outdoor steps to the right of a Catholic church called La Iglesia Parroquial de la Santísima Trinidad at Plaza Mayor, the town's main square. Adjacent to this church, we had had a lovely dinner at a restaurant just off the square, with a view of a brilliant sunset, with chirpy locals, excited tourists and occasional horse-riders passing by. From soon after dinner till way past midnight, Cuban music and salsa dancing energized the entire neighbourhood. Here was a microcosm of Cuban flavour in world culture.

In all of us is a bit of Cuba.

The outdoor 'stage' and 'dance floor' are on a landing about halfway up the steps. Chairs are placed wherever there is space, some perched precariously on a narrow step. But no one seemed to care about anything other than to get a good view, a cool drink and a slice of the hot action. Musicians all dressed in matching red tops, on keyboard, guitar, drums and trumpet, provided rousing rhythms non-stop. The soloist, a middle-aged man, had a voice that was as rich and deep as it was hoarse, his only dance move a constant slight shuffle. His role was to provide the vocals; he did not need to be good-looking, he just needed to sound good. Each evening, it is the patrons who are the main performers. And there is never any lack of volunteers.

Every night, no shortage of guests take turns to show off their moves. This night was no exception. Women in bareback dresses with swinging skirts sashayed with flair, the sinews of their arching backs glistening with little beads of perspiration, while the men were just as good dancers but somehow seemed to have to exert themselves just a little more in comparison, simply to keep up. The tourists were beginning to feel that this was going to be a night to remember. The locals have never had any doubt.

The electric vibes of Cuba can also be felt at other venues such as the Buena Vista Social Club, which features classic Cuban live music

Feel the beat: Salsa in full swing at Casa de la Música (House of Music) in Trinidad.

over dinner, and the Fabrica de Arte Cubano, a trendy, avant-garde nightspot with a bar, live performances and contemporary visual arts exhibitions. Such merriment was made all the more memorable and poignant for me because of the setting and all the history that wraps around it. Cuba, essentially, is an island of isolation and invasion. It is the kind of isolation that only a scourge as devastating as Covid-19, with its lockdowns and closed borders, has made many people around the world more able now to empathize with it. With channels of external trade choked off for much of its recent history, Cuba has had to strive for material survival for a very long time. Stoic steadfastness, seasoned with the solace of salsa, have offered some comfort, snatched from the pendulum swings of fortune. In recent years, a brighter future had seemed to be dawning, with the opening up that was beginning during the Obama administration in the USA. From 2016, in the Trump administration, these relaxations had begun to be reversed, but now, under President Joseph Biden, things might change yet again.

Here was a bubble long before the pandemic. With widely available public internet access introduced only from late 2018, sometime after

Daddy cool: Street musicians in Trinidad, even at lunchtime, in the noonday sun.

our visit in early summer, Cuba has for a long time had to make sense of its own place in the world more or less on its own. Social distancing from the rest of the world was very much part of life. Foreign powers have always fought and schemed to exert control over this highly strategic maritime node in the waters between the Pacific and Atlantic oceans. But through it all, Cuba's ubiquitous cultural affinity with people around the globe enables it to hold a special place in the hearts of active and would-be travellers worldwide, despite the isolation of the island in almost every sense of the word, and the many invasions into its territory throughout history.

For years, many people have unknowingly expressed the Cuban spirit, as it infused world music. Clubbers everywhere would know the 2017 salsa-inspired hit single 'Havana' by Cuban-American singer Camilla Cabello, which has entranced a younger generation with the sex appeal of Cuba, expressed through well-known, hypnotic lyrics such as this line: 'Half of my heart is in Havana, ooh-na-na (ay, ay)'. As for people who grew up in an earlier era, they would be more familiar with the song 'Guantanamera', which occupies a spot on the most-requested lists of karaoke rooms across the globe.

Many a karaoke wannabe has performed this number with gusto, not realizing that the chorus sings the praises of a guajira ('peasant woman' or 'country girl' in Spanish) from the south-eastern city of Guantánamo (the Guantanamera of the title). Guantánamo, of course, later became a household name for a different reason in the post-911 world as the name of a bay on the coast where the US still maintains a naval base and detention camp. In late 2018, the base marked the 115th anniversary of when the US first occupied it in 1903. This was just after the war with the Spanish that led to the Europeans surrendering, signing the Treaty of Paris and then handing over Cuba, Puerto Rico, the Philippines and Guam to the US.

People around the world who have hummed the tune and imbibed the spirit of the song 'Guantanamera' probably know Cuba's political history only vaguely and superficially. What they know much better are the Cuban rhythms that have stirred the hearts of people everywhere, from Havana to Singapore. Few people think about the irony that the song 'Guantanamera' was first adapted by the American folk singer

Roarin' 50s: American vintage cars cruising past El Capitolio (National Capitol Building) and Hotel Inglaterra in Havana, and in front of the City Hall in Cienfuegos, a city on the south coast.

and campaigner Pete Seeger into an international peace anthem in the early 1960s, a time when the Cuban missile crisis was casting a shadow over the world, much like the North Korean nuclear crisis has done more recently. 'Guantanamera' was a Cuban-born song with an easy-to-remember chorus that audiences could sing along to, to express international solidarity.

Global Cuba: The Cuba Libre café and bar in Cecil Street in Singapore—complete with Cuban flag and map as dance floor backdrop centrepiece—shows the country brand's international success.

To be sure, the cultural linkages work both ways, such as when Cubans perform international music. At the casa particular (inn) where we were staying in Trinidad, it was great to hear Leonard Cohen's classic song 'Hallelujah' being sung in Spanish by two female singers accompanied by a male guitarist in the courtyard during happy hour. Certainly, music brings people together like few other things can.

The evolving landscape of accommodation in Cuba reflects broader changes in society. Casa particulares are available almost everywhere in the tourist areas of Cuba. A relaxation of regulations by the government in recent years allowed ordinary Cubans to convert their private homes into something like Airbnb accommodation, totally Cuban-run. The casa in Havana where we stayed the first night was the proprietor's home, with two bedrooms available for guests, in contrast to the modern high-rise boutique hotel where we stayed at the end of our visit to Cuba. The casa in Trinidad was a two-storey building with a large courtyard and several rooms, and even had a restaurant and bar. The casa in Viñales was a single-storey terrace house with three bedrooms. The proprietors lived next door and came over in the mornings to prepare breakfast. Viñales is a quaint town in the western part of Cuba, which gives a totally different vibe from the city lifestyle of Havana. Its colourful colonial-era wooden houses are set in Viñales Valley, with vast stretches of surrounding farmland and a backdrop of limestone hills known as mogotes.

Cuba Libre everywhere

The universal familiarity with the song 'Guantanamera' reflects the tremendous international presence and brand recognition of Cuban culture. The Cuba Libre (originally a cry for political liberation meaning 'free Cuba') bar and café in Singapore is an example of the power of brand Cuba. It expanded from its first outlet in the nightlife district of Clarke Quay, a lively staple of the bar scene there since 2006, to open another bar and café in Singapore's Chinatown, and also in Kuala Lumpur, Malaysia.

In Singapore, the Cuba Libre brand of cafés and bars offers an authentic slice of Cuba, where patrons can immerse themselves in the spirit of Cuba in Cuban music and dancing—of course, lubricated with the help of drinks like the one after which the establishment is named. The grasp of the political symbolism and significance of 'Cuba Libre' by the bar's patrons might be a tad superficial mostly, with the notion of 'freedom' associated more with hanging loose on a good night out. Nonetheless, the brand affinity is universal and deep.

The Cuban spirit is captured also in the visual arts across the country, reflecting the power of art to rejuvenate communities and inspire people. In Havana, one place that throbs with this energy is Fusterlandia, the nickname given to Jaimanitas, a seaside town west of downtown Havana, and named after artist José Fuster, whose house can be visited here. Once a poor neighbourhood, the whole precinct was reinvented with the multi-coloured mosaics of Fuster. His house is full of his artistic creations, from the swimming pool to knick-knacks in every nook and cranny, and his artworks also adorn many houses in the neighbourhood.

Beneath the surface gloss, the back story of the very idea of Cuba Libre is full of historical meaning. The drink Cuba Libre is said to have been created around 1900 to celebrate Cuba's independence from Spanish occupation. It perhaps sums up a major chapter of the island's history that the drink's main ingredient is Coca-Cola, the quintessential American beverage. The local integration is complete when the Coke is mixed with Cuban rum—Havana Club rum in places where Cuban products can be exported to, or, if not, usually Bacardi rum elsewhere.

Indeed, this concoction, like most countries, has its own founding myth—the drink is said to have been created by accident. A 1960s Bacardi advertising campaign describes how American soldiers, who had helped Cuban rebels defeat the Spanish colonialists at the beginning of the twentieth century, were in a Havana bar and wanted to spice up their Coca-Cola drinks by going for the obvious upgrade—throwing in some Bacardi rum and also adding lime. They called the drink 'Cuba Libre!', a rallying call of victory to celebrate yet another successful military intervention. Ironically, then, a glass of Cuba Libre celebrates a liberation that is actually only partial, with the incursion of one foreign power replacing another.

Another irony of freedom is that America's influence has left one of its best legacies in a symbol of mobility with style. One of the best

Cuban whimsy: Fusterlandia, a precinct with extensive public art installations by the maverick artist José Fuster, whose colourful mosaics evoke revolutionary triumph ('el triunfo', written on the arch over this gate) and a dreamy wonderland.

tourist experiences in Havana is to take a ride in an American vintage car, usually a convertible Chevrolet or a similar make from the 1950s. Gliding through town, with the top down and the balmy weather reinvigorated by the breeze, Cuba becomes an even more romantic place. A ride in one of these cars of yesteryear is an experience that can be cherished not only on the streets of Havana on a city tour, but also on a hired drive to explore the countryside, in and around other towns. They certainly do not make cars like they used to. It makes you wonder why automobile companies stopped making cars beautiful.

Here is another aspect of American soft power that has captured the consciousness of all things Cuban. These vintage cars make for such attractive sights on the streets of the Cuban capital. Often in neon cartoon-like hues such as shocking pink or lime green, they would look more at home on a Hollywood movie set or on the artificial streets of a Universal Studios theme park. These iconic symbols of Havana in particular, and Cuba in general, are unmistakable reminders of Cuba's relationship with the US, however ambivalent this is.

A glass of Cuba Libre celebrates a liberation that is actually only partial, with the incursion of one foreign power replacing another.

One wonders how Cuban car mechanics still manage to find spare parts and the know-how to maintain these vehicles. If these cars from so many decades ago can still be maintained and run so well today, why are they not seen in other cities? Why can't they still be made commercially, since there obviously is a market for them? Putting it down to the typical Cuban resourcefulness, it is not surprising that they would somehow manage to get the job done to maintain these vehicles. From the many stories our guide Aneil told us of how he often has to improvise to get around the constraints of life in Cuba, such as material shortages and other limitations and regulations, clearly, Cubans get on with life by making the best of what they have or can acquire. How much more resourceful would they have had to be since this pandemic surfaced?

The most amazing example of Aneil's resourcefulness was when we visited the Buena Vista coffee plantation at Las Terrazas, a small community and nature reserve on the road from Havana to Viñales, in

Reminders of revolution: A poster calling on then-President Batista to leave office, at Havana's Museum of the Revolution, and a billboard with the slogan 'Fidel forever' on the road to Viñales.

the Sierra del Rosario mountains, a UNESCO Biosphere Reserve. As we were chatting on our walk, I voiced my love for wildlife photography and how I wished to photograph Cuba's national bird if we had the chance. A few minutes later, Aneil said he would walk on ahead into the forest to try to spot the Cuban trogon (nicknamed the tocororo by Cubans, a name taken from the sound of its call). True enough, he did manage to find one in a tree near us. We were stumped by the coincidence that we

Lest we forget: Revolution Plaza at Santa Clara, and a bookshop in Havana.

happened to be in a place where Cuban trogons could be found at that very moment, and also by his uncanny ability to locate one so quickly.

Later, at the Soroa Orchid Garden in Viñales, we spotted another Cuban trogon, relaxed enough in the tree above us for more, and clearer, photographs to be taken. What an extremely beautiful and apt national bird this is, as its feathers uniquely sport the red, white and blue colours of the national flag—a perfect brand icon. These two tocororo sightings were simply unforgettable.

The revolution lives on

In the making of nation brands, the eternal tension between self-sufficiency and global connectivity captures the position of islands such as Cuba in the global context. Just as Singapore will always be of geostrategic interest and importance in the South China Sea, Cuba holds a similar maritime value in the Caribbean. The two islands allow whoever commands its surrounding sea channels to hold open the economic lifelines that course through the adjacent waters, as well as to maintain the security of the region.

Cuba, however, has suffered so much more than other similarly coveted states. Cuba's struggle for independence dates back to 1492, when the explorer Christopher Columbus 'discovered' the Americas and claimed Cuba for Spain. Thus began a process of domination that irrevocably altered the indigenous cultures of Latin America. This 'cultural continent' includes thirteen dependencies and twenty countries where Romance languages such as Spanish, Portuguese and French are spoken. Geographically, the region stretches from the northern border of Mexico to the southern tip of South America and the Caribbean, and includes a few of the countries we visited on this journey.

In Cuba, Spanish rule lasted around four centuries until the Spanish–American War of 1898, when American military assistance helped Cuban rebels end Spanish colonization after thirty years of rebellion. However, the planned celebrations did not happen because the Americans took control of the territory, and have, to this day, continued to occupy the area surrounding Guantánamo Bay, which was leased in perpetuity to the US in the 1903 treaty.

For any society, being at the centre of such longstanding struggle can bring strife to everyday life. While the rhythms that Cuban musicians create are always rousing—how could it be otherwise, given the infectious cadence of the maracas and other Cuban musical instruments—sometimes, their everyday body language might betray something else going on deeper down.

National hero: Bust of poet and philosopher José Martí in the town square at Viñales.

Whether in the lunchtime entertainment on the second storey of a hip bistro in Havana, or the early evening troupe in a more traditional Spanish-style restaurant in Trinidad at sunset, there are visible signs that the level of enthusiasm of these musicians is sometimes not operating at maximum capacity, especially during low season. This dip in energy, one could surmise, might be due to factors like residual fatigue from the daily routine, but it might also indicate stoic resignation, or some other undercurrents in the larger picture.

This observation can perhaps be taken to be emblematic of the country as a whole. The labour market has, to say the least, some measure of under-employment. The situation must have deteriorated further during this pandemic. As an illustration, some tour guides, who are trained engineers with advanced degrees, would probably rather be practising what they went to university to learn. But the reality is that most engineering jobs are in the government sector, where the salaries are low by any standard, especially because most salaries are paid in national

Vital link: Tourists tapping the internet connectivity at the public Wi-Fi hotspot at Casa de la Musica in Trinidad.

pesos, in Cuba's dual currency system. One national peso is equivalent to one quarter of a convertible peso, the other form of currency that can be exchanged for US dollars.

By contrast, the tourism industry—one of the country's main interfaces with the outside world and primary sources of foreign currency—pays much better. At least, working as tour guides puts into practice a skill that is not commonplace, with a solid barrier to entry: not just fluency in English, but enough of a linguistic command to be able to hold a conversation that often requires them to dip into a wider vocabulary, to share their musings on the state of current affairs and even facets of the meaning of life.

Connecting with the world

Some of these conversations can be surprising, given the breadth of perspective they reveal when set against the constraints on access to information about the outside world. Before late 2018, when the Cuban government set about allowing internet access across the country, most Cubans and foreign visitors could go online only by tapping the limited connectivity at and around Wi-Fi hotspots in public parks, town squares or government-run hotels.

Internet access was gained at that time by using scratch cards with passwords that enabled two hours of access through a government internet service provider. These navigation cards were sold in government offices, which always had predictably long queues. An informal economy and secondary market for these cards ran on reselling them for about double their value—so, say, three or more US dollars instead of two for those who did not want to spend time queuing. As recently as 2007, it was illegal to purchase a computer in Cuba. In 2009, a man faced imprisonment for fifteen years for setting up a Wi-Fi network because this was deemed a threat to 'the integrity of the state'.

As a result, parks and squares were always crowded with people, all peering intently into their phones and trying to do what they needed to do online as quickly as possible to make their access cards last longer. Any younger visitor from the digital native generation would feel discomfort and withdrawal symptoms in no time.

Various methods are deployed by enterprising Cubans to circumvent censorship and other forms of internet control. With the opening to online access since late 2018, the government has shifted its stance more and more towards opening up in content regulation. What is interesting is how Cuban people could have been so relaxed despite their isolation, and generally be so well-informed about the state of the world outside even with such limitations by First World norms.

For most Cubans, life includes deprivation in some form or other. Food, drink and other essential supplies are curbed by severe shortages and limits from the trade embargos maintained through most of the island's history. At the same time, there are the restrictions of law and order, as well as residual structural rigidities in all aspects of society, as a legacy of state communism. Residents still visit government shops to collect their rations of assigned portions of basic foodstuffs and other necessities. In socialist societies, a layer of realism holds everything together, with the glue of resignation from too many uneventful yesterdays, moistened by patient hope for a brighter tomorrow.

A layer of realism holds everything together, with the glue of resignation from too many uneventful yesterdays, moistened by patient hope for a brighter tomorrow.

There are marks everywhere in Cuba of the resilience and resourcefulness of the Cuban people. Quite apart from the hardships of daily life, Cubans seem happy to make the most of what life has to offer, including embracing foreign powers, if doing this might mean more economic opportunity. In addition to the dominance of American soft power, there is also the odd sign of ambivalence from Cubans themselves, as seen in those who wear clothes emblazoned with the US flag, despite all the troubles that have been visited on the island by US forces through the decades of interference with the sovereignty and security of this place.

A powerful strand of the Cuban spirit is the inspiration that is drawn from the nationalist narratives, heroes and icons that are impossible to avoid or ignore. Images and slogans abound everywhere to mark the 1959 revolution led by communist revolutionary Fidel

Castro to overthrow the then-President Fulgencio Batista, replacing an authoritarian government with a revolutionary socialist state. Just as, if not more, prominent are the memorials to the national hero José Martí, an important figure in Latin American literature.

Spirit of the land: One of many art galleries celebrating indigenous culture, in the southern city of Cienfuegos.

Martí is honoured everywhere in Cuba, from the international airport in Havana bearing his name to busts in many of the most prominent places, including in the main square of Viñales. Martí was a Cuban writer and political activist who became a symbol of Cuba's bid for independence against Spain in the nineteenth century. One of his poems from the book *Versos Sencillos* (*Simple Verses*) was adapted for the lyrics of the song 'Guantanamera', which is also considered the definitive patriotic song of Cuba. Martí's poem reads as follows:

> I have a white rose to tend
> In July as in January;
> I give it to the true friend
> Who offers his frank hand to me.
> And for the cruel one whose blows
> Break the heart by which I live,
> Thistle nor thorn do I give:
> For him, too, I have a white rose.

Former US President Barack Obama quoted from this poem in his landmark speech on his vision for the future of Cuba during his visit to the island in 2016. As the first sitting US President to visit Cuba since Calvin Coolidge in 1928, Obama made a valiant attempt to bury the last remnants of the Cold War with the old Soviet Union, as Cuba served as a proxy in the Caribbean for the conflict between the two superpowers. The poem's sentiment of remaining open to extending an olive branch to one's opponents—'thistle nor thorn do I give: For him, too, I have a white rose'—is echoed later in the stance towards one's political adversaries made famous by his wife Michelle in her speeches during the 2016 US presidential election: 'If they go low, we go high'.

Nonetheless, for Cuba, truly, there is nothing like art and literature—in the poetry of Martí and others—to give the deepest sense of soul, of what matters most to a people, what is at the heart of their politics and desire for their own future. It is a testament to the power of literature that the Museo Ernest Hemingway—a villa on a hill in San Francisco de Paula, 15 km southeast of Havana, where he lived continuously for twenty years—is said to be the most popular museum

Old man and the scene: Bronze statue of Hemingway at the very spot where visitors can imagine him downing his daiquiris in his favourite corner of El Floridita fish restaurant and cocktail bar in Havana.

in Cuba for foreign visitors, understandably so, as most tourists are from the US. This is so even though visitors can only peer into the house from the outside, through the building's French windows, as a special permit is needed to enter the premises.

Finca Vigía, or Lookout Farm, is where the American writer made his home from 1939 to 1960, where he wrote seven books, including the 1952 novella *The Old Man And The Sea*, his last major work. The story is of Santiago, an ageing Cuban fisherman who snags a giant blue marlin off the coast of Cuba, only to have it devoured by white sharks. Defeated, he returns home with the fish's skeleton attached to the boat. A tale of perseverance against formidable odds, in a maritime setting that is quintessential Cuba, this story has some very Cuban qualities indeed—most of all, fortitude amidst adversity. Thus, Hemingway is one of Cuba's foreign brand ambassadors par excellence.

Thankfully, memories of crucial aspects of history are preserved in Cuba's temples of heritage. Being able to visit a museum, or read a book of poems, enables a people to stay attuned to the emotional and spiritual foundations for the most precious parts of their own being. It is as if art can teleport them in their imaginations back in time, to remind themselves of what has brought them to today. Art can also transport them forward, to imagine what their shared visions might be capable of holding in future.

Cuba Libre: A perfect accompaniment to yet another brilliant Havana sunset.

Memories of crucial aspects of history are preserved in Cuba's temples of heritage. Being able to visit a museum, or read a book of poems, enables a people to stay attuned to the emotional and spiritual foundations for the most precious parts of their own being.

In this vein, visiting the Museum of the Revolution in Havana is a must, to get a big-picture sense of the struggle for freedom that produced today's society and system of government. Other important sites include the memorial in Havana to Castro and Che Guevara, Castro's second-in-command, and the museum at Revolution Plaza in Santa Clara, on the road to Trinidad, where the exhibits are more personal. These reminders of how long the Cuban people have fought to gain more agency in life are sobering reminders to people everywhere, especially citizens of small states, of how precious freedom and independence truly are. These are also precious peeks into the human exertions, through tribulation and triumph, that have enabled this island to overcome the unceasing challenges of isolation and invasion.

Panama

Betwixt two oceans

The haiku poem and ink sketch in this haiga artwork seek to sum up what it means to be such a significant piece of land: An isthmus between two great bodies of water, as well as the link between the worlds on each side—and thus, forever a part of both halves. This will always be the cornerstone of Panama's country brand—with global connection at its core, where even the inevitability of geography was jettisoned with the ingenuity of building and sustaining its Canal.

PANAMA

Isthmus identity

being between is
to be bound with beyond, both
flanks of this axis

許木松
2020

Only connect: The Panama Canal transformed the globe when it linked two oceans and two worlds.

Isthmus of change

'Only connect', wrote the English writer E. M. Forster (1879–1970). Panama made this happen for the world. For some countries, there is no question about their paramount brand icon. Just mention Panama and straight away people think of the Panama Canal—that wonder of human engineering that cut through land to link two oceans, thereby connecting 160 countries and 1,700 ports across the planet. When this tremendous feat of human engineering was officially opened in 1914, it changed the world forever. This artificial waterway enabled ships to avoid the lengthy, hazardous Cape Horn route around the southernmost tip of South America via the Drake Passage or Strait of Magellan. Straddling an isthmus, this land bridge was formed approximately 2.8 million years ago connecting the North American and South American continents. The building of the canal changed the course of history, and also transformed the people of Panama's sense of themselves.

Here is where those with the mood, money and means of transport can wake up beside the Caribbean Sea leading into the Atlantic Ocean and later have lunch overlooking the Pacific Ocean.

Here is where those with the mood, money and means of transport can wake up beside the Caribbean Sea leading into the Atlantic Ocean and later have lunch overlooking the Pacific Ocean, for the two oceans are just a couple of hours apart, on a fast drive, across the isthmus' 'skinniest' sections. The narrowest of all the Central American countries, Panama is 59.5 km wide at its narrowest point and 177 km at its broadest. From the top of Volcán Baru, Panama's tallest mountain at 3,475 metres, in the province of Chiriquí, on a clear day, you can see two oceans while standing in one spot.

The evocative power of this isthmus is captured in the poem 'My Country' by Ricardo Miró Denis (1883–1940), the Panamanian writer considered to be the most noteworthy poet of his country. Here are the first, second and last stanzas, translated by Tom Pow:

Such a small country, spread on an isthmus
where the sky is clearer, the sun brighter;

all your music echoes within me, like the sea
in the small cell of the conch.
Yet again, there are times I feel dread
when I don't see the way back to you . . .
Perhaps I'd never have known such love,
if destiny hadn't carried me over the sea.
. . .
Such a small country, all of you will fit
beneath the shadow of our flag: perhaps
you were so pretty, to ensure I'd carry you
everywhere in my heart!

The poet's yearning to embrace all the beauty and meaning of his homeland—within the 'small cell of the conch', perhaps a metaphor for his earthly body and his very being—should have been comforted in its

Game changer: The Panama Canal changed the course of human history—literally and figuratively.

ambition, given the comparatively small size of the land. Yet, at the same time, he knows that to 'fit' everything would be possible only in the abstract, in his heart. Miró's words also touch on nationalism—'beneath the shadow of our flag'—referencing the young age of Panama from its independence from Colombia in 1903.

Looking out over the canal, you can feel Miró's passion. There are places on Earth that you journey to just to stand for a breathtaking while and soak in their history, to visualize the imagination and exertions that were invested at those very spots to affect the course of human destiny. This is one of them, for sure. The thought of the many lives that had been changed by this game-changing piece of engineering was simply mind-boggling.

Naturally, for a place known for feats of engineering, the more 'mechanical' parts of the tourism infrastructure are also up to scratch. These include the access and exit routes for the tour buses, the ticketing system, information kiosks, refreshment amenities, and, of course, the merchandise shop. But it is the observation deck, opposite the Miraflores Locks, that this place is all about. Here, one can look up and down the canal, absorbing the spirit of this isthmus. If your stay in Panama City is long enough for your visit to the canal to coincide with a ship passing through, you can witness this wonder of shipping. Every year, more than 15,000 vessels make the journey through the waterway, which can take around eight hours.

The Panama Canal's 82-km waterway is an engineer's dream come true. Canal locks at each end lift ships 26 metres above sea level up to Gatun Lake, an artificial water body created to reduce the amount of excavation work required for the canal, and then lower the ships at the other end. A third, wider lane of locks, which took nine years to construct, was completed in 2016, when the expanded canal began commercial operations. The new locks allow transit of larger, neo-Panamax ships, capable of handling more cargo.

Panama's immense geostrategic importance made it the focus of attention of foreign powers. In 1513, the Spanish explorer Vasco Nunez de Balbao discovered that the isthmus of Panama was but a slim land bridge separating the Atlantic and Pacific oceans. Not long after, Panama became part of the Spanish Empire, the first modern global empire, for

European heritage: French Square in honour of those who worked on the Panama Canal and the Spanish San Francisco de Asis church in Casco Viejo (Old Quarter).

over 300 years (1513–1821), with the Spanish exploring overland routes to cross the oceans. Then followed a few decades of political turmoil with new political alliances forming in South America, before Panama became part of the new Republic of Colombia.

Colombia, France and later the US controlled the territory surrounding the canal during its construction. After completing in 1869 the Suez Canal connecting the Mediterranean Sea to the Red Sea, the French company led by Ferdinand de Lesseps believed that an apparently similar project to connect the Atlantic and Pacific oceans could be carried out with little difficulty. In 1881, de Lesseps' company broke ground in Panama to attempt to build such a crossing. His plan was to complete the project at sea level, thus eliminating the need for locks, but the tough geography and geology had other ideas. Frustrated by this, the French team recruited the engineer Gustave Eiffel, of Eiffel Tower fame, to design and create the lock system needed for the canal. However, plagued with poor planning, engineering issues and tropical diseases including yellow fever and malaria, thousands of labourers met their untimely end, and the project was halted.

Let's do the twist: Panama City's 'screwdriver' skyscraper F&F Tower, reflecting a recent trend in international building design.

The US had shown great interest in a trans-American canal for economic and military reasons and had originally considered one in Nicaragua. But they were persuaded otherwise by Phillippe-Jean Banua-Varilla, a French engineer who had worked in de Lesseps' company, and began lobbying American lawmakers in the late 1890s to buy over

Reaching out: Panama, like other small states such as Singapore, is looking to land reclamation to acquire more space for the city to grow.

the French canal assets earlier abandoned in Panama. He eventually convinced a number of them that Nicaragua had dangerous volcanoes, making Panama the safer choice. However, even today, the Nicaraguan option is still on the table more than 100 years on, with a Chinese company announcing that it has struck a US$40 billion deal to begin construction of a waterway to rival the Panama Canal, although this project drifted into a state of limbo, and is now delayed even further by the pandemic.

> For some countries, there is no question about their paramount brand icon. Just mention Panama, and straight away, people think of the Panama Canal.

In 1902, the US Congress authorized the purchase of the former French assets, but there was one small problem. Panama was at that time part of Colombia, which refused to ratify the agreement. Unbelievably, in a twist of fate that was like a godsend for the US, the Panamanians actually revolted and won independence from Colombia. A year later, the US supported Panama's secession from Colombia and secured a sovereign zone in which to build the Panama Canal. Under the Hay-Banua-Varilla Treaty, the zone was to be an American protectorate for perpetuity. The US would shell out some US$375 million to construct the canal, which included a US$10 million payment to Panama as a condition of the 1903 treaty, and US$40 million to buy the French assets.

The US took over the project in 1904 and opened the canal on 15 August 1914. It continued to control the canal and the surrounding Panama Canal Zone until the 1977 Torrijos–Carter Treaties provided for handover to Panama. In 1989 the US invaded Panama to depose a former ally, military ruler Manuel Noriega, over his repressive rule and use of the country as a centre for drug trafficking. After a period of joint American–Panamanian control, in 1999, the canal was taken over by the Panamanian government. It is now managed and operated by the government-owned Panama Canal Authority. Annual traffic rose from about 1,000 ships in 1914, when the canal opened, to just over 14,000 vessels in 2008, the steady-state level that has been sustained since.

A paragon of modernity

Like other small states such as Singapore, Panama is looking to increase its living and operating space. Land reclamation is one major option. Man-made islands and other reclamation projects are extending the shoreline and opening up new opportunities for waterfront properties—and even here, there is indirect impact from the Panama Canal. These reclamation projects were made possible by using soil from the excavations to build the canal. The reclaimed land is also expanding the Panamanians' sense of their world in more ways than one. Flying into Panama City, the shoreline scenery is striking.

Panama City's downtown shoreline resembles somewhat the southern coastline of Singapore as seen from a plane flying into Changi Airport. I did not think of home more at any other time during the sixty-eight days of this trip. There is a cluster of skyscrapers near the shore, seeking to maximize the high-value ambience of sea views, proximity to sea breezes and access to the coast, with waterfront walkways and stretches of beach at some spots. This indicates the attractiveness of Panama as a destination for various profiles of residents, including those looking for holiday or retirement homes, such as foreigners from the US and elsewhere. Unlike in Singapore, one difference in Panama City is the height of the buildings close to the water's edge, instead of a more stepped, or cascaded, profile. The latter would allow for more apartment or office units to enjoy views of the sea.

Panama is a paragon of modernity, in many ways. This is its best-known place brand attribute.

Panama is a paragon of modernity, in many ways. This is its best-known place brand attribute. Its capital, Panama City, is considered by some as the only true First World city in Central America. The beautifully maintained Pan-American Highway runs the breadth of the country, making travel easy. High-speed internet and cell phone coverage are top-notch, as are the power, air and water quality.

Being connected with the rest of the world is what the Panama Canal offers in concrete terms, literally. The air hub that is Tocumen

International Airport makes it easy to fly from Panama to nearly anywhere in the world, often with no layovers. All the country brand attributes related to modernity are alluded to in the tourism slogan that Panama had used previously: 'Where the world meets'.

This, indeed, is where the world meets. For expats from the US, Panama is also convenient because the US currency is used as legal tender. The Panamanian balboa has always been tied to the US dollar, and Panama does not even print its own paper currency, although it has its own coins. This is another aspect of the extent to which nations in Central America leverage off the global dominance of the US in many aspects, especially in economics, finance and trade. For anyone interested in analysing the world's 'global cities', Panama would certainly make for an interesting case study. If, in pre-Covid-19 times, Singapore was Asia's poster child of globalization, then Panama was surely Central America's representative.

Hats off to the Panama hat

Alongside the Canal, the Panama hat—with its signature black band around its inner circumference—has instant brand recall for Panama. It is probably one of the most striking cases of the power of branding in the retail sector. The signature straw hat actually originated in Ecuador, but very few people know this, instead mistakenly believing it to be a product of Panama—so much so that Ecuador has given up trying to 'correct' this common misconception.

As far back as the sixteenth century, the making of Panama hats emerged as a cottage industry within the coastal communities of Ecuador, Panama's neighbour to the south. As many as fifteen people are involved in the process of making a Panama hat, which begins with harvesting the shoots from the toquilla palm grown in the coastal regions of Ecuador. The hats' original function was to shield workers on the Panama Canal from the sun's harsh ultra-violet rays, so as to boost their productivity and get the job done faster. But, thanks to one particular event, the vital, game-changing importance of the hat as a brand ambassador item for Panama was realized in 1906 when US President Theodore Roosevelt was photographed wearing a Panama hat whilst visiting the construction site of the canal to inspect its progress.

'Beauties of Panama': That is this shop's name in Spanish, and the Panama hat certainly qualifies.

From the moment the media splashed photos of Roosevelt wearing the Panama hat, it quickly became the epitome of elegance and sophistication amongst the social elite, elevating the image of the hat to iconic status. In so doing, the American president catapulted to a global icon what had originated as something with a much humbler function.

These accessories came to be named 'Panama hats' originally on account of their point of trade rather than their country of origin. The hats were traded through the isthmus of Panama, and the name Panama came to be attached indelibly to this headgear. This was a coup for Panama, one of the best examples of a brand takeover.

Panama served as the gateway to the eastern US and Europe, where the popularity of this fashion statement grew to what it is today. Well-dressed men and women in the US and Europe were captivated by these finely woven hats. The Panama hat became part of the de rigueur attire for Europeans and Americans for their explorations of new world domains and territories. Its popularity grew and this trademark look found its way into popular media and culture through films, as personified by the silver screen's heroes and adventurers.

The hats came to be named 'Panama' hats originally on account of their point of trade rather than their country of origin. This was a coup for Panama, one of the best examples of a brand takeover.

As for me, I was sorely tempted to buy a Panama hat in Panama City, but, after some agonizing, decided against it for two main reasons. The first was a practical one, out of typical Singaporean pragmatism—being fully aware that there were still four more countries to visit on this sixty-eight-day journey, and knowing that I had quite a few luggage items to manage on the move, including a backpack, camera and so on in addition to the suitcases, having to look after a Panama hat and protect it from being crushed would be an added challenge. On a flight, squeezing the hat into an overhead compartment would squash it out of shape in no time. Also, because the hat has no chinstrap, it was liable to fly off with a breeze of any strength at any time whenever it is worn outdoors.

The other reason was that I already owned an 'original' Panama hat that I had bought some ten years ago on a trip to Ecuador. That hat had a band in the yellow, blue and red colours of Ecuador's national flag, instead of the Panama hat's classic black. Hence, I decided to let this opportunity to buy a Panama hat in Panama go and have somewhat regretted it ever since. Every time I am reminded of this, I console

The sloth and the tree frog: Species that link Panama's biodiversity with that of its neighbours in Central America, such as Costa Rica.

myself that had I bought it, it would surely have been damaged in transit. Maybe, maybe not . . .

In mid-2020, I happened one day to pass a shop in Singapore that specializes in selling Panama hats imported from Ecuador. So I have one now, in classic 'Sean Connery style', and no longer have those feelings of regret. Instead, in addition to all the memories from this trip, a tangible piece of Panama is now with me every day.

Biodiversity and a sanctuary for species

A surprising fact about Panama, given its relatively small land area, is that it has the largest rainforest in the western hemisphere outside the Amazon basin in South America. Just a bit larger in land area than Ireland, Panama has more species of birds, amphibians and animals (if insects are included) than the US and Canada combined.

When Panama emerged from the sea through plate tectonics three million years ago, the unusual geography of the isthmus meant that

Tribute to the wild: The Biomuseo (Biodiversity Museum) designed by Frank Gehry, seen here in the distance from our hotel balcony.

Panama, in addition to transforming forever the course of maritime trade, also affected the biodiversity of the world. As a barrier between the Atlantic and the Pacific oceans, the formation of Panama changed the wind patterns, water currents, salinity and other factors affecting climate, causing the regions and the wildlife touched by the oceans to evolve in different ecological directions. The climatic development around the isthmus further created a gulf stream effect that defines rainfall amounts, temperatures, tides and climates for the whole region.

The plate tectonics that produced Panama also created a sanctuary for species including the unique pygmy sloths. Pygmy sloths are a dwarf version of the brown-throated three-fingered sloths found on mainland Panama, but this species is endemic to the remote Isla Escudo de Veraguas. This tiny island, which formed only 9,000 years ago, is located 17.6 km off the coast of mainland Panama and covers just 4.3 sq km. Pygmy sloths are considered one of the world's most critically endangered mammals, with the last official population count identifying fewer than eighty individuals. Very little is known with scientific rigour about these little sloths.

The pygmy sloths were first described as a distinct species in 2001 based on morphological differences in body size (they are reportedly 40 per cent smaller than the mainland species), but due to a lack of research, there is still no genetic information on the status of these animals. Isla Escudo is very difficult to get to, and much of the mixed forest covering the interior of the island remains unexplored. The pygmy sloths are most commonly found inhabiting in, and feeding from, red mangrove thickets. In addition to the Isla Escudo, there

On the go: Panama's city developments create an ever-changing skyline across from the ancient town, seen here from the main waterfront promenade.

are known to be four other nearby islands in the Bocas del Toro archipelago that support three-fingered sloths smaller in size due to their confinement on an island.

To celebrate Panama's biodiversity, the Biomuseo in Panama City is a museum focused on the natural history of Panama. Located on the Amador Causeway, it was designed by the renowned architect Frank Gehry, his first design for Latin America. It was hoped that this museum would have a similarly transformative effect on Panama City as Gehry's Guggenheim design had on Bilbao, which rejuvenated and placed the city on the map as an important architectural destination. But the circumstances of Panama, including its relatively greater distance from Europe and Asia, meant that the impact has not been quite the same. Hence, the gap in brand awareness between the two museums is quite considerable and would not be easy to bridge.

A museum like the Biomuseo in a place like Panama might seem like a random project for Frank Gehry, until one learns that his wife is Panamanian. The angular, asymmetrical, primary colour design of the building was inspired by a few important aspects: the clothing of Panama's indigenous Kuna and Emberá peoples, the tin roofs and colourful buildings of local Panamanian architecture, the angles and 'artistic chaos' of the shipping containers that pass through the Panama Canal and the geological upheaval of the isthmus that originally created Panama.

Panama, the Biomuseo proclaims, became a 'bridge of life' (the title of its permanent exhibition) and a fountain of biodiversity. Even today, the land connection between the two continents means that the movement and evolution of flora and fauna are ongoing in both directions, to the north and south. One effect that has been observed is that southern species have found it harder to adapt to colder northern climates, compared with northern species that moved south to warmer habitats. Anthony G. Coates, a Smithsonian scientist who worked on the Biomuseo, has written that 'today over 60 per cent of mammal species in South America are derived from northern forms. However, only three southern species survived in North America (the opossum, armadillo and hedgehog) although some species, such as the capybara, are still migrating northwards today.'

Football fever: At Panama City Airport, patriotic pastries (with a Spanish name meaning 'supporting Panama') are on sale next to a sports bar screening the opening ceremony of the 2018 World Cup in Russia.

The concept for the Biomuseo, including a six-acre biodiversity park, was formulated in 1999 as the US was handing over the canal's administration to Panama. The museum was conceived as a tourist attraction, an educational facility and a part of the country's new identity and brand. Panama provided the land and loans, while the Amador Foundation raised private money to build and operate the Biomuseo.

People of Panama: Clockwise from top—The different faces of human diversity, including a mural remembering the 1925 indigenous revolution, a hip Vietnamese-Japanese restaurant at Plaza Herrera Casco Viejo and a roadside arch built by the local Chinese community.

Open to the rest of the world

Panama's diversity, most of all, is in its people. There is something about places that are close to the sea, and therefore more open to the rest of the world, that makes them more attractive. Here, there is a cosmopolitan vibe everywhere, which gels with the juxtaposition of the old town and the new city that one can see in one view from a stroll along the city's main waterfront promenade, which we did one evening before dinner. The fancy fusion cuisine meal reinvented old Panamanian recipes, using traditional ingredients in innovative ways for an international palette, thus exemplifying Panama City's orientation to the world.

The people of Panama have a reputation for having big hearts and an even bigger zest for life. They are welcoming to foreigners, who in turn feel safe here. Increasing numbers of North Americans, Europeans and others are moving here and contributing to the burgeoning economy. Impressed by Panama's cosmopolitan vibes and obvious connectivity to the world, I decided to try an experiment to test the extent of an average Panama City resident's awareness of current affairs. The day we took our walking tour in Panama City was the day after the North Korea–United States Singapore Summit in Singapore, where US President Donald Trump met North Korean Chairman Kim Jong Un. I asked our guide Ramos whether he knew about the event and what he thought of it, wondering if he might have more of an opinion of Singapore than before, for being selected as a location to host the event.

Panama's diversity, most of all, is in its people. There is something about places that are close to the sea, and therefore more open to the rest of the world, that makes them more attractive.

Some commentators had made much of Singapore playing the role of a 'Switzerland of the East' by hosting this Summit. Ramos did not think much beyond it being a significant world event. Singapore, I suppose, cannot expect too much credit for being essentially a meeting venue. I guess most Panamanians would not have made too much of the occasion. After all, like most people anywhere, their focus in life is on doing their job, doing what they can to extend a welcome to visitors and then going back to enjoying life. This is why nation brand-building is such hard work. What this shows is that just as it would take a lot of effort by Singapore to further enhance its standing in the world, the same applies to other territories like Panama.

As in many other societies, Panama's ethnic diversity has a broad pattern. First, there is an indigenous segment of the population, who suffered under the colonialism of decades past and still carries deep memories of subjugation and indignity. Second, descendants of the waves of early immigrants under the colonial era would probably have formed the majority of the political, economic and social elites. Third, immigrants who helped build the nation in its earlier years are, for the most part, now an assimilated and accepted minority. And fourth, more recent immigrants have made the place even more diverse, trendy and forward-looking.

A wall mural that we saw marked the ninetieth anniversary in 2015 of the 1925 Guna Revolution, in which the Guna Indian tribe fought the ruling Panamanian authorities, who were attempting to force the Indians to adopt Hispanic culture by military coercion. During this revolution, the Guna Yala territory seceded and operated as the short-lived Republic of Tule. Following mediation by the US, the Guna people were reunited with Panama. The Gunas, with the support of the Panamanian government, created an autonomous territory called the Guna Yala district for the indigenous inhabitants, which they would rule themselves. Today, they are partially integrated into Panamanian society, but clearly, as this mural shows, some old memories have not totally receded.

Elsewhere, there are other marks of ethnic diversity. Chinese arches and other architectural features reflect the presence of a thriving Chinese community. The first Chinese immigrated to Panama from southern China in the nineteenth century to help build the Panama Railroad. There followed several waves of immigrants, adding up to about 4 per cent today of a total population of just over four million, believed to be the largest Chinese population segment in a Central American nation. Restaurants serving Japanese and Vietnamese cuisine reflect more recent waves of immigrants from other parts of Asia, as well as the more cosmopolitan tastes of a more modern Panama.

What has united the people of Panama are shared national passions, such as sport, through support for the national team. We had the chance to see a bit of this during our visit. At Panama City Airport, shops were selling 'patriotic pastries', decorated with football-themed icing and other features, while the sports bars were screening the World Cup tournament opening ceremonies and matches. Panama was one of the two smallest nations participating in the World Cup Finals for the first time, along with Iceland (with fewer than 350,000 citizens). Sport unites people like few other aspects of life can. This is true all the more for small states.

Participating in the football World Cup despite its physical tininess is one example of how Panama is an inspiration to all small countries, that limitations of size should never be allowed to curb the potential of a people. What matters most is what value a place can add to the rest of the world. For Panama, a principal role is to be an isthmus that connects the whole globe, and this has been the foundation for so much achievement in the past, and now, for aspiration for the future. Panama's connection to the world will lead the way forward once again, after Covid-19 and beyond.

Mexico

Seeking firmer ground

The haiku poem and ink sketch in this haiga artwork place Mexico's Aztec treasures, such as the Teotihuacán complex, within the sphere of a very human impulse that is shared from ancient times till today—to make sense of the universe and acknowledge the larger forces of nature. The scale of the surviving heritage places this at the heart of the country brand: a major piece of civilization with much more in its foundations to be revealed and appreciated.

MEXICO

Shared purpose

it's only human
to reach for the sun, for a
lasting monument

許木松
2020

Story of resilience: Mexico's history is a narrative of an ancient civilization that succumbed to deterioration, struggled against colonialism, struck out with independence, and still strives today for self-determination.

Biggest city of the Americas: From Teotihuacán to today

On this journey, among the thirteen countries were five that we had been to before, including the UAE, Spain, Japan and Taiwan. For Mexico, there were two main reasons for us to visit again. The first was that there was still so much of Mexico that we had not seen. We had been to Mexico twice before, both times as extensions of trips that also included the US—first, in 1994 to the beach resort city of Cancun, from where we took a day trip to visit the ancient Mayan ruins of Chichen Itza and Tulum, and then, in 2004 to Cozumel as part of a Caribbean cruise, which started from Florida and also stopped at Jamaica and the Grand Cayman Islands.

As for the second reason: We had never been to Mexico City before, and one thing we always try to do is to include at least a brief visit to the capital of any country we are visiting. The capital is where key elements of a society are always located—including the seat of government, which indicates the priorities of governance; the most important memorials of history and museums containing the most significant artefacts; and glimpses into how the elite organize and lead their lives.

Size usually matters, when it comes to brand power and awareness. Mexico City was, in ancient times, the biggest city in the Americas.

Mexico has a few other compelling reasons to demand more attention. Size usually matters, when it comes to brand power and awareness. Mexico City was, in ancient times, the biggest city in the Americas, and with a population estimated at 125,000 or more, it was at least the sixth-largest city in the world during its epoch. Till today, it dominates the region with its economic heft. The Mexico City metropolitan area has a population of more than twenty-one million people, making it the largest metropolitan area in the western hemisphere, the fifth largest in the world, and the world's largest Spanish-speaking city.

Despite its scale, however, one of Mexico's most troublesome challenges of nation brand-building is the extent to which brand perceptions of it can be framed—and thus, coloured—by being viewed through the lens of its northern neighbour, the US, with its dominance of international media and pop culture. What Mexico has to do is to boost its nation brand on its own terms.

Mexico City itself has some major sights, and it is also an excellent place from which to visit Teotihuacán, just 40 km to the northeast. It is this proximity to Mexico City that helped enhance Teotihuacán's standing as one of the world's busiest ancient sites. Designated a UNESCO World Heritage Site in 1987, Teotihuacán is the most-visited archaeological site in Mexico, receiving more than four million visitors a year.

Teotihuacán is an ancient Mesoamerican city, home to an important culture that predated the Aztecs, who flourished in central Mexico in the post-Classic period from 1300 to 1521. Some people consider Teotihuacán more impressive than Chichen Itza, even though Chichen Itza is home to El Castillo, or the Pyramid of Kukulcan, an ancient Mayan ruin that some consider is to Mexico what the Eiffel Tower is to France as a national icon. Having been to Chichen Itza before, we had to make the comparison for ourselves.

Although it is a subject of debate whether Teotihuacán was the centre of a state empire, its influence throughout Mesoamerica is well documented. Scholars have suggested that Teotihuacán was a multi-ethnic state, possibly including the Nahua, Otomi or Totonac ethnic groups. This would make it one of the earliest models of multiculturalism, exemplifying a pragmatic way of coexistence and cooperation to boost the collective chances of survival.

The earliest buildings at Teotihuacán date back to around 200 BCE. The largest pyramid, the Pyramid of the Sun, was completed by 100 CE. The name Teotihuacán was given by the Nahuatl-speaking Aztecs centuries after the fall of the city around 550 CE. The term has been glossed as 'the birthplace of the gods', reflecting creation myths that were said to occur in Teotihuacán. The Aztecs believed that the gods created the universe at this site.

When visiting this site, an early start to the day is recommended, to avoid the crowds that typically throng this location later and throughout the day. Our guide Carlos sounded very experienced and had obviously worked out this drill many times before. He had the demeanour of a university professor and delivered his briefings to the English-speaking

Aztec relics: Mexico's heritage has many impressive artefacts, and some of the best are housed in the National Museum of Anthropology in Mexico City.

half of our group as if we were on an archaeological field trip. At some locations, the steps of the ancient pyramids could be used as seating for tourists, just like in an ancient amphitheatre. We all took our seats like undergraduates on excursion, to listen to his engaging commentary. As with many such tours to ancient sites, the shared details of interest at the

Sun and moon: The morning sun, rendered hazy by overcast skies, looks like the moon in this reflection in a puddle near the Pyramid of the Moon at Teotihuacán.

site bring to life some of the history of what must have been a thriving civilization.

Teotihuacán's main thoroughfare encourages comparisons with other ancient civilizations with regard to beliefs surrounding the supernatural and the afterlife. The main corridor through the whole complex is called Avenida de los Muertos (Avenue of the Dead), and the pyramids here, like their counterparts elsewhere from Cairo to Cambodia, are essentially a way to come to terms with the uncertainties of what lies beyond this human existence. This aspect captures a fascinating facet of the Mexican psyche—the way in which Mexico mulls over mortality. As the Mexican Nobel Laureate poet Octavio Paz (1914–98) writes in his 1950 book *The Labyrinth Of Solitude:*

> 'To live is to be separated from what we were in order to approach what we are going to be in the mysterious future. Solitude is the profoundest fact of the human condition. Man is the only being who knows he is alone, and the only one who seeks out another.'

What coping with solitude entails includes the multi-dimensional complexity (the 'labyrinth') of Mexico's history. The course of human development contains so much the pain of seeking 'separation from what we were' and the inevitable contestation in marshalling people and resources to go along with how to get to 'what we are going to be'.

Aside from the pyramids, Teotihuacán is also anthropologically significant. Its complex, multi-family residential compounds and those murals that have been well-preserved give a sense of the structures of family and government that established and sustained societal order and fostered harmony in this settlement. This was also a centre of an export economy, as Teotihuacán is known to be the source of fine tools made from the volcanic glass called obsidian, which has been found throughout Mesoamerica. The city is thought to have been established around 100 BCE, with major monuments continuously under construction until about 250 CE. Scholars still debate what really happened at Teotihuacán. Previously, experts had thought that invaders attacked the city in the seventh or eighth centuries, sacking and burning it. But more recent evidence seems to show that no traces of foreign invasion are visible at the site. It is now believed that the burning was centred on major civic structures along the Avenue of the Dead and the dwellings associated primarily with the ruling class. The sculptures inside palatial structures were also shattered.

Top of the pile: Teotihuacán is the most-visited of Mexico's magnificent ancient ruins, with millions drawn to the mystique of the Pyramids of the Sun and Moon.

Some think this suggests that the destruction came from an internal uprising. This could perhaps have been a protest against inequality, like the unrest that surfaced across the world from around 2015 into the rest of the 2010s. These uprisings included what was dubbed the 'Latin American Spring', in reference to the protests in Chile, Ecuador and other places in South America. These incidents were compared with the Arab Spring in the early 2010s, the series of anti-government protests in a few parts of the Islamic world, starting with Tunisia and spreading to other countries including Egypt and Syria.

Evidence of population declines at Teotihuacán beginning around the sixth century lends some support to the internal unrest hypothesis. The eventual downfall of Teotihuacán has been correlated to lengthy droughts related to climate change. This theory of ecological deterioration is supported by archaeological remains that show a rise in the percentage of juvenile skeletons with evidence of malnutrition during the sixth century, suggesting that famine is most likely one of the more possible reasons for the waning of Teotihuacán's fortunes. The majority of their food came from agriculture: They grew things such as maize, beans, amaranth, green tomatoes and pumpkins, but their harvest was not nearly sufficient to feed a population as big as is believed to have lived in Teotihuacán.

> Theories of the collapse of an entire civilization echo the forces at work across the globe today—climate change, inequality and civilian protest. If Teotihuacán could fall, then anything that exists today could also fall just as easily.

Sudden destruction as in Teotihuacán was a common fate for Mesoamerican city-states of the Classic period (300–950 CE). There is also another theory that the collapse of Teotihuacán was caused by its agriculture being devastated by the 535 CE eruption of the Ilopango volcano in El Salvador, to the south beyond Guatemala. These theories of the end of an entire civilization carry echoes of some of the forces at work across the globe today—climate change, inequality and civilian protest. If Teotihuacán could fall, then anything that exists today could also fall just as easily.

Life goes on: Hill-slope shanty towns on the road between Teotihuacán and Mexico City are typical of rapid urban development around the capital, where many Mexicans market replicas of ancient cultural heritage as well as more modern items for their livelihood.

Heritage buried in the heart of the city

Fascinated with Teotihuacán, we were keen to learn more. And further exploration into Mexican history and heritage can definitely be conducted right in Mexico City itself. The best place to start your visit is in the centre of the capital city, at the site of the Templo Mayor (Spanish for 'Greater Temple'), which was the main temple of the Mexican people in their capital city of Tenochtitlan (Mexico City today).

The Templo Mayor was discovered in 1978 while workers were digging to lay underground electric cables, and what a find that was—this place is, in every sense, where it all started for this slice of civilization. Perhaps the city's stray dogs were on to something—this particular spot was popularly known as the 'island of the dogs', because it was slightly higher than the rest of the neighbourhood, and in times of flooding, wandering canines in the street would gather there.

The architectural style of the Templo Mayor belongs to the late post-Classic period of Mesoamerica (950–1521). The temple was called the Huēyi Teōcalli in the Nahuatl language. It was dedicated simultaneously to Huitzilopochtli, the god of war, and Tlaloc, god of rain and agriculture. Like many ancient civilizations, the people attributed divine actors and powers to the forces of nature that they could not understand otherwise.

The scale of Mexico's ancient settlements is mirrored in the Mexico City of today. Mexico's capital is both the oldest capital city in the Americas and one of two founded by Native Americans, the other being Quito in Ecuador. On the basis of GDP, Mexico City alone would rank as the thirtieth largest economy in the world. This city had been the capital of the Aztec empire, and in the colonial era, Mexico City became the capital of New Spain. The viceroy of Mexico or vice-king lived in the vice-regal palace on Zócalo (the main square, today called the Plaza de la Constitución).

As with almost all of Latin America, reminders of so many centuries of Spanish rule still stand firm today. The Mexico City Metropolitan Cathedral of the Assumption of the Most Blessed Virgin Mary into the Heavens is the seat of the Catholic Archdiocese of Mexico. It was built by the Spaniards over the ruins of the main Aztec temple. In this way, it is an architectural metaphor of imperialism, one that embodies, literally,

Sacred ground: The Recinto Sagrado (Sacred Precinct) in Mexico City harbours many relics of the culture's ancient origins, right in the heart of the capital.

the layering on and imposition of colonialism over ancient heritage. A police officer we saw standing in front of the cathedral added yet another modern layer of irony, a further overlay of another level of even more recent authority.

After a decade of war, Mexico's sovereignty from Spain was established with the Declaration of Independence of the Mexican Empire on 27 September 1821. Unrest followed for the next several decades, as different factions fought for control of Mexico.

Imperial overlay: Mexico City's Metropolitan Cathedral, built by the Spanish over the ruins of the main ancient Aztec temple.

Other landmarks of Mexico City are more recent. The most recognizable icon of Mexico City is the golden Angel of Independence, built in 1910 on the wide, elegant avenue Paseo de la Reforma, which was modelled, by order of Emperor Maximilian of Mexico, after the Champs-Élysées boulevard in Paris, France. This was also a blueprint for many other such grand pieces of national infrastructure across Central and South America. This avenue, the Americas' oldest known major roadway, was designed to connect the National Palace (the seat of government) with the Castle of Chapultepec, the imperial residence. Today, this avenue is a key artery of the financial district in which the Mexican Stock Exchange and several corporate headquarters are located.

A whole history in one artwork

Very often, the quintessence of a people is captured in art; sometimes in one artwork alone. There is one sight in Mexico City that I would put at the top of the must-see list: *The History Of Mexico* murals in and around the stairwells of the National Palace. This masterpiece was executed between 1929 and 1935 by Diego Rivera (1886–1957), one of Mexico's most famous artists. His full name, remarkably, was: Diego María de la Concepción Juan Nepomuceno Estanislao de la Rivera y Barrientos Acosta y Rodríguez.

As a window into a foreign culture, typically, the visual arts may not have the scope for depth, elaboration and honesty that only literature can provide, but they do offer the most evocative, inspiring vistas. The subject of Rivera's magnificent murals transcends even words, as it is unbelievably ambitious in scope and detail: Mexico's entire history from ancient times to the present. They depict the many struggles of the common Mexican people for liberty and self-determination—their unending fights against the Spanish and the French, as well as the dictators who controlled the country at different times.

Rivera was a communist radical, known for his criticism of authority and foreign domination. Even though his art was sponsored by the Mexican government, he was allowed to continue to rebel against authority of any kind. His paintings were highly political and rejected capitalism but glorified the indigenous cultures of Mexico including the Aztecs and the Zapotecs. After studying at the San Carlos Academy, Rivera went to Europe, where he learned about Cubism and Renaissance frescoes. He returned to Mexico and applied his ideas about Cubism and other European and Mexican art traditions to the Mexican mural movement. He was married to Frida Kahlo, also an internationally famous Mexican artist.

Canvas of community: The *History Of Mexico* murals at the National Palace in Mexico City by Diego Rivera tell the history of Mexico across the centuries.

The Mexican government had commissioned Rivera and other famous Mexican artists including Orozco and Siqueiros to paint scenes about Mexican history. In August 1929, Rivera began painting his huge murals in the large stairways and stairwells of the National Palace, the centre of the Mexican government and nation. Rivera was hired to portray the native people in a better light, and to criticize the Spaniards. The main aim was to celebrate the Mexican Revolution, the overthrow of the Porfirio Diaz dictatorship, and the new government.

The National Palace is located on the Zocalo, the central plaza in Mexico City from where Moctezuma, the Aztec king, had ruled Mexico before the arrival of the Spanish. To view the *History Of Mexico* murals, we needed to register and get a lanyard, just like visitors to government buildings typically have to.

The subject of Diego Rivera's magnificent murals . . . is unbelievably ambitious in scope and detail: Mexico's entire history from ancient times to the present.

The *History Of Mexico* murals are huge, some as big as 70 metres by 9 metres. The North Wall section displays the richness of the ancient Aztec culture, especially the people's traditional costumes. The sun was the centre of the world in Aztec divinity, as exemplified by the Pyramid of the Sun at Teotihuacán. Rivera depicts the everyday life of the Aztecs, including the volcanoes around the Valley of Mexico, corn and other main crops, artisans weaving and making pottery, mothers carrying babies, art and artisan activities including painting on scrolls and calligraphy. The section also illustrates the Aztecs' religion, including the worship of snakes and jaguars. Aztec rituals are also shown, in which men consume alcoholic beverages made from cacti and dance in honour of the Sun God.

The South Wall depicts the future of Mexico, with imagery of progress and prosperity. Scenes show factories, the Soviet flag (Rivera and the government at the time held socialist beliefs), workers, Karl Marx and the Communist Manifesto. These visuals promoting the Revolution take up a minority of the space. Rivera's main subject is the deep cultural heritage of the Native Americans and the greatness of Mexican civilization before the Spanish conquest. The murals illustrate

Faces of Mexico City: Just as Rivera's *History Of Mexico* murals capture the struggles of ordinary people, today's 'street characters' include a police officer with breakfast in hand, and an organillero ('organ grinder') street musician cranking out 1930s classics.

the true culture of Mexico: bright colours suggesting hope and the beauty of the country and its people, as well as the more sombre hues of its old customs and rich traditions.

The West Wall is the central part of the murals and summarizes the violent but beautiful history of Mexico as a series of conflicts of rebellion and revolution against oppression. Common Mexicans and Indians revolt against the Spanish and French and various dictators. Scenes depicted include the Spanish conqueror Cortes defeating the Aztecs and other Indians, and Spanish priests during the Inquisition trying to obliterate the sacred books and religious images of the Indians. Presented here is the origin story of a key emblem of Mexico's nation brand. The most important picture in Rivera's main mural is of a large eagle with a serpent in its mouth, which represents the Aztec culture. This is now the main image that symbolizes modern Mexico and is found at the centre of the national flag.

It is possible to see the Mexico of today as a real-life enactment of what has been so artfully captured in paint by Rivera in terms of the dynamics between the centres of power and the people.

Rivera's masterpiece gives awesome visual representation to what the poet Octavio Paz called 'the labyrinth of solitude', the immense struggles of human life made more tumultuous by the nagging doubts that the only progress is to be found alone. Taking one step back from the murals, it is possible to see the Mexico of today as a real-life enactment of what has been so artfully captured in paint by Rivera in terms of the dynamics between the centres of power and the people, which are at work in most societies throughout history. The current authority structures are modern, but those in charge are perhaps just as liable to succumb to the temptations of hegemonic dominance.

What has changed is the agency of the ordinary person—some who may be more set to repeat the errors of history, while others, taking in the wisdom of the past, can adopt a more enlightened approach to the world. Art such as Rivera's murals has distilled the essence of Mexico's heritage, but effort must still be made to extract the lessons of human endeavour. In archaeological terms, Mexico is still finding out more about itself, whether through the excavations at the ancient archaeological sites of Templo Mayor or at Teotihuacán. Similarly, journeys of self-discovery are still open within Diego Rivera's murals. This, too, is about seeking firmer ground.

Modern monuments: Mexico City's Palacio de Bellas Artes (Palace of Fine Arts) and El Ángel (Angel of Independence), built in 1910 to commemorate the centennial of the beginning of Mexico's War of Independence.

Labyrinth of solitude: As Mexican poet Octavio Paz writes, human struggle becomes more tumultuous if progress is only to be found alone.

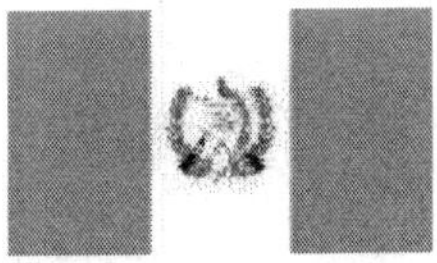

Guatemala

Tradition amidst turmoil

The haiku poem and ink sketch in this haiga artwork re-enact the scene of millions of photos, the Santa Catalina Arch in Antigua, which is a fitting brand icon of Guatemala in more ways than one. The volcano backdrop, where clouds often swirl and ash sometimes spews, echoes the turbulence of eras past, and also resembles stage curtains over the arch itself, while the cobblestoned ground recalls the country's rough, uneven but solid path through history.

GUATEMALA
Passion play
cloud dance of fire
drapes this stage of centuries:
cobbled desire
許木松
2020

Volcanic past: Antigua, for 200 years the colonial capital of Guatemala until an earthquake in 1773, is a symbol of a history of uneasy tussle.

Antigua: Once the heart of an ancient kingdom

Originally, our itinerary for this journey had us spending a few more days in Mexico. Then we thought: why not fit in one more destination? There were enough days in the buffer on our schedule to do that. So, we looked at the world map once more and decided to add Guatemala, Mexico's neighbour. Looking back, we are so glad we did. We did not know anyone who had visited Guatemala, although Guatemala is actually an easy trip from Mexico. Places that not many people have been to are somehow all the more exciting. Besides, my friend from graduate school, Luz, lives there, and I hoped we might be able to meet her, if she was in town.

What also helped sealed the deal for us to add Guatemala to the itinerary was seeing online the iconic image of Antigua Guatemala ('Old Guatemala' in Spanish), commonly referred to as la Antigua or just Antigua. This UNESCO World Heritage site was the colonial capital of Guatemala, after the arrival of the Spanish in the sixteenth century. Clearly, the architects back then had oriented the town such that the Santa Catalina arch, Antigua's main sight, was set precisely against a backdrop of the most impressive volcano in the vicinity, thus creating a stunning scene. Just being able to stand in this spot alone, if nothing else, would make the visit to Guatemala worthwhile.

Strolling through Antigua's cobbled streets is literally a walk back in time to Spanish colonial days.

For sure, there are very few locations more exotic than this place in South America. The makers of the 1935 film *The New Adventures Of Tarzan* obviously thought so too. The movie, starring Herman Brix, was shot on location in Guatemala, taking advantage of help from then-president Jorge Ubico. The filming was done at the town of Chichicastenango, for the scenes of a native town where the explorers first met, and the ruins of San Francisco Church in Antigua were used for location shooting, as the 'Green Goddess temple' dedicated to this Guatemalan idol in the movie.

Guatemala once formed the core of the Mayan civilization, which extended across Mesoamerica, before the Spanish conquerors took over

Arch of history: The Santa Catalina Arch is the pre-eminent symbol of Antigua, and of Guatemala too.

this land and stamped their mark on it. Named Santiago de los Caballeros by the Spanish, what is present-day Guatemala was the capital of the kingdom of Guatemala, that covered a much larger area. This extensive territory stretched from the modern state of Chiapas in central Mexico through the current states of Guatemala, Belize, El Salvador, Honduras, Nicaragua and northern Costa Rica. Strolling through Antigua's cobbled

Still on fire: Volcán de Fuego (Volcano of Fire)—emitting steam after an eruption a few days ago—makes for a dramatic backdrop to Antigua's 'skyline', in real life and on canvas.

streets is literally a walk back in time to Spanish colonial days. This long history does not have top-of-mind recall for most travellers. But it is, in fact, a rich reservoir of culture and heritage for the nation brand of Guatemala to be built on.

It was intensely gratifying when we eventually set foot on the spot where millions of photos had been taken and were able to gaze up at the Santa Catalina Arch with the awesome natural backdrop of the Volcán

Centre of celebration: The baroque La Merced church—*the* place in Antigua to hold a wedding, one of which we gatecrashed for a bit on arrival in this charming ancient town.

de Fuego (Volcano of Fire, some 3,763 metres high). What made it even more dramatic was the knowledge that the volcano had erupted just a few days before our visit. We found out later that the eruption in June 2018 resulted in about 160 deaths and at least 300 injuries, more than 250 missing persons and some 3,000 residents being evacuated from the surrounding areas, as well as the closure of La Aurora International Airport for one day because of the ash cloud.

Volcán de Fuego is famous for being almost constantly active at a low level. Small gas and ash eruptions occur every fifteen to twenty minutes, but larger eruptions are rare. That day, the volcano was still spewing clouds of steam and smoke. This, together with the low-lying clouds, obscured the view of the peak of the volcano above the Santa Catalina Arch. This was the only minor minus point of being there at that time, but it was an unforgettable sight nonetheless. Volcán de Fuego is one of the three large volcanoes dominating Antigua's horizon. To the south of the city is the Volcán de Agua (Volcano of Water), some 3,766 metres high, dormant and possibly extinct. To the west of the city is Acatenango, some 3,976 metres high, which last erupted in 1972.

Antigua is renowned for its preserved Spanish Baroque-influenced architecture as well as a number of ruins of colonial churches and

Riot of colour: The charm of the indoor and outdoor markets of Antigua.

buildings, some dating back to when Guatemala was conquered by Spanish conquistador Pedro de Alvarado in 1524. A 1773 earthquake ended Antigua's 200-year reign as Guatemala's colonial capital, and many buildings were restored thereafter. Guatemala became a republic in 1839 after the United Provinces of Central America collapsed. Notable architectural examples that still stand today include the baroque La Merced church. It is an integral part of the city's famous celebration of Semana Santa, a holy week with parades and rituals.

Originally a monastery, which opened in 1767, the La Merced church is a monument to the ingenuity of earlier engineers—the building was constructed at a low height, designed to better withstand earthquakes, based on the experiences of the tremors of 1751. The arches and columns were also made wider to cope with the quakes. It has stood the test of time remarkably well. On the morning that we visited, a wedding was about to take place. The guests were decked in their finest apparel, and the church was decorated with colourful flowers and buntings. We were happy to be amongst the beneficiaries of this living display of culture and tradition.

Antigua is such an ancient place, yet it is only now becoming a fast-growing tourist destination in Guatemala. A big advantage is being close to Guatemala City—a mere forty-five-minute drive away—and it is much calmer and safer, with more tourist-oriented activities. It is a favourite getaway for local Guatemalans, who throng Antigua on the weekends. It is possible to take buses from Antigua on to many other parts of Guatemala. Transport links are, of course, more plentiful from the hub of Guatemala City, but from Antigua, many travel agencies offer shuttle rides to the main tourist places: Monterrico beach, Atitlan Lake, and the towns of Coban, Lanquín (Semuc Champey) or Tikal.

Thanks to its volcanic landscape, Antigua has the ideal climate for growing coffee, and the beans produced here are considered amongst the best in Latin America. Coffee aficionados can visit a coffee plantation, where farmers will proudly show off their crops and the coffee-making process. Many coffee shops scattered throughout town allow you to sample the different varieties of Antiguan coffee. Amongst other attractions, Antigua is known for its chocolate makers. Guatemala is known as the birthplace of chocolate, and Antigua has a few cacao

Palette of society: Old and new Guatemala City, as seen in the brightly hued homes on the outskirts, in contrast to the more subdued colour of the corporate and entertainment district 'Zona Viva'.

museums where you can learn how the Mayans made chocolate—a produce they considered 'the food of the gods'.

Shopping for handicrafts is also fun, as Guatemalan village women are deservedly known for their dexterity with needle and thread. Antigua's vibrant markets—both indoors and outdoors—are stocked with colourful handicrafts, pretty woven garments and travel souvenirs.

Perfect accompaniment: Traditional Guatemalan marimba music goes very well with traditional meals at heritage restaurants.

Other items, some intricately sewn, include ponchos, hats, tablecloths, wallets and decorative bowls. Many visual artists make Antigua their base. Their favourite subject is, of course, the Santa Catalina Arch with Volcán de Fuego in the background.

Back to reality: Guatemala City

The road back to Guatemala City offers a glimpse of the everyday life of the majority of Guatemalans in the brightly painted homes on the hillsides. The sight is a colourful reminder of the palette of segments that make the whole canvas of society. These tenements seemed perched midway, on a halfway point between the colonial heritage of a place like Antigua and the urban sprawl and hectic modernity of the modern capital. This is a bit like Guatemala itself, poised between tradition and modernity.

Guatemala City is, in so many ways, a world apart from Antigua. The colonial charm and calm of Antigua, an ancient capital of a massive region, exists in snippets in some quarters of Guatemala City, the modern capital of a populous state. But, in the main, the places of interest are dotted amongst a society still coming to terms with many facets of administration, organization and cohesion. Visitors are cautioned against moving around on their own, in a place that used to rank amongst the cities with the world's highest rates of murder and crime overall.

The lingering social issues are sustained by society's lesser elements, but they have also called forth the best from the residents. Our friend Luz, whom we met for dinner, works in the non-profit sector to advocate for the betterment of marginalized groups in Guatemalan society. Her decades of self-sacrifice and dedication to her cause reflect the resilience and resourcefulness of the people of Guatemala.

One thread that runs through the more troubled episodes of Guatemala's history is that life is often more about strife than success, more hewing than harvest. The Covid-19 pandemic has driven home this reality all the more. As the Guatemalan Nobel Laureate poet Miguel Ángel Asturias (1899–1974) wrote in his poem 'The Fortune':

> We were made that way,
> Made to scatter
> Seeds in the furrow
> And stars in the ocean

Tea dance: Residents of Guatemala City dance the afternoon away at a cultural festival in a neighbourhood park.

Having to 'scatter seeds in the furrow' is perfectly understandable, part of the necessary labour of agriculture and other work necessary to sustain human existence. But throwing 'stars in the ocean' would be a much more uncertain investment, which, unfortunately, so much more of life requires.

Today, amongst Guatemala's lingering social issues, one of particular concern is juvenile crime. A recent trend is that more people have turned to buying cars and driving themselves around so as to avoid using public transport because this is where many youth gangs operate. At a time when some cities around the world are working to reduce car ownership to help save the environment, here, cars can be a matter of life and death. The onset of Covid-19 will certainly pose additional societal challenges, making the need for greater social resilience all the more urgent. Some social commentators theorize that the predisposition towards violence in Guatemalan society is perhaps a vestige of history, stemming from the thirty-year-long civil war that ended in 1996. If this is true, it can only be countered by substantial and concerted investment in renewing a shared sense of community, and this invariably must seek its richest sustenance from a deep sense of place and identity. Drawing on Guatemala's rich heritage would be an excellent place to start.

Inner city issues are the bane, for sure, of many big cities. And Guatemala is a larger and more significant nation than many people realize. It is the most populated country in Central America, with more than seventeen million citizens and residents. The preserved heritage and surrounding natural scenery of Antigua make it a central location from which to visit other tourist areas in Guatemala and Central America. Cruise ships that dock at Guatemalan ports offer trips to Antigua from both the Pacific and Atlantic oceans.

Renewing a shared sense of community invariably draws its richest sustenance from a deep sense of place and identity. Drawing on Guatemala's rich heritage would be an excellent place to start.

The sense is that most visitors complete their trips before they get the chance to gain a deeper appreciation of Guatemala's heritage and culture, and even those who want to are unfortunately hindered from doing so by concerns such as those over public safety. In the city, tours with a guide are fine, which is what we did. But even then, whenever we disembarked, we were told to stick close together and make our stops focused and brief.

Nonetheless, we felt perfectly safe. With our guide Ramos, we ventured out to get a feel of the city on a combination of vehicle recce

Place in the sun: Guatemalan women and children have a day out in the grounds of the Antigua Guatemala Cathedral, where a reconstructed fountain is a popular local gathering spot.

and walking tour. He always made sure that our driver would be ready to pick us up at the end point of our walk through any particular location. We certainly enjoyed getting a sense of everyday life in Guatemala City. For example, Parque Central (Central Park) is where everyone in Guatemala goes out on the town. It is a social gathering place for people of all ages and from all walks of life.

A chance snapshot in the square turned out to be one of my favourites from the whole trip. This spontaneous composition included the national flag both on a flagpole in the centre of the square and also in the colours of the football jersey of the boy who is the photo's main subject, held in the arms of a man, presumably his father. Wrapping around the boy, and also, figuratively, the photo itself, is the social aspect of his doting parents. Meanwhile, in the background is another symbol of Guatemalan nationalism in the National Palace of Culture, built in 1943 as the seat of government of Guatemala and now a museum. This location marked the starting point of all the roads in the republic and has a spot in the centre known as Kilometro Cero (Zero Kilometre). This place marking the historical origin of modern Guatemala is an apt location to place an image for the country's renewed sense of its future, captured in the expression of cheerful curiosity on the boy's face.

Aside from places such as Parque Central, international visitors would feel most at home walking around on their own in Zona Viva (Lively Zone), the name used to refer to a part of Zone 10 in Guatemala City, east of the impressive Avenida la Reforma. This avenue runs through the periphery of Zona Viva and is flanked by many monuments to Guatemala's history. It was originally known as Boulevard 30 de Junio (30 June), to commemorate the liberal reforms launched on that day in 1871. Then-president Reina Barrios had wanted to showcase Guatemala's economic boom for the 1897 Central American Expo. Drawing inspiration from the Champs-Elysées in Paris, he commissioned several sculptures and fountains from foreign artists. He also commissioned the Palacio de la Reforma, a French-style palace at the southern end of the boulevard, which served as a museum but was destroyed in the 1917–18 earthquakes.

Like other roads in Latin America, Avenida la Reforma was inspired by the Champs-Elysées boulevard in Paris, which shows how the Spanish,

Central park: A family on a day out in Parque Central (Central Park), the social heart of Guatemala City, with the National Palace of Culture in the background. The image of Mickey Mouse on the backdrop of the tent reflects the reach and impact of the soft power of the US.

like many people of all eras, were fans of French architecture and urban planning. This zone is considered a more affluent district with a high density of bars, restaurants, hotels and nightspots. In addition, a great number of shopping malls and office buildings make this area a bustling business centre of Guatemala City. Multiple well-known hotel chains run some of the finest accommodations in the country here. In past decades, dozens of new high-rise buildings have sprung up. This is also the most liveable part of the Guatemalan capital, a district that remains quite a green area with many parkways.

We were here for a very pleasant, and safe, stroll on a Sunday morning. We had just had breakfast at a packed café that was showing the live telecast of the World Cup football match between Mexico and Germany. Aside from passing vehicles, our only companions were joggers, brisk walkers and dog-walkers. Some might call this precinct a 'bubble'—Zona Viva, the fancy part of town, set apart from the rest of the city and kept safer than other areas with stepped-up security and the social norm of most residents typically not venturing too far from their spheres of everyday life, by force of habit. It is a lifestyle that would seem to make it easier to adjust to something like a pandemic lockdown.

In the same way, Antigua is another kind of bubble, a well-preserved relic of a vital history that, till today, is a centre of cultural and economic vibrancy as it draws travellers from across the globe and provides a venue of weekend relief for city-dwellers from the modern capital city. As for Guatemala City, it is also a bubble in some ways, in a region periodically beset by political and social upheaval. Overall, in the context of our visit to Guatemala, the very idea of a 'bubble' also sums up one key aspect of the state of any modern, still developing, society: Even if unsettling forces may be adjacent, the ballast of heritage and tradition can offer sanctuary and solace. Such spiritual resources will be vital for eventual recovery from the setbacks of Covid-19.

French chic: Avenida la Reforma in Guatemala City, inspired by the Champs-Elysées boulevard in Paris.

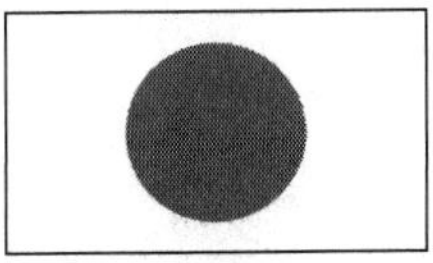

Japan

At Southeast Asia's edge

The haiku poem and ink sketch in this haiga artwork celebrate Shuri Castle's place as the brand icon of Okinawa and as a perfect symbol in many ways for this unique corner of Japan, at its interface with Southeast Asia and the world outside. The castle's history of re-emergence from the ashes of destruction epitomizes the country's own key brand attributes of age-old resilience and spirit, alongside an unstinting advance towards modernity.

JAPAN
Rising sun
a fort's fortitude
is not just its standing, but
how well it's rescued
許木松
2020

Gateway to the world: An entrance to Shuri Castle, emblem of Okinawa's unique culture, symbolizes Japan's boundary here with Southeast Asia.

Okinawa: Japan's 'most un-Japanese place'

A few stabs of pain pierced my heart when I saw online the news about the fire at Shuri Castle in Naha, Okinawa, on 31 October 2019. The main courtyard structures of the 500-year-old World Heritage Site were reduced to ash and rubble in a blaze, suspected to have been sparked by an electrical short-circuit, although initial official investigations ended without identifying a definite cause. Shuri Castle is as important to Okinawa as Edinburgh Castle is to the Scottish city, or Buckingham Palace to London. To me, it is the spiritual heart of Okinawa, this exceptional, 'most un-Japanese place' in Japan, which I was glad I had the chance to visit not once, but twice.

Both visits now hold an even more special place in my heart, poignantly and yes, painfully exemplifying the Japanese concept of ichigo ichie—the awareness that every single experience in life should be cherished and treasured because it might very well turn out to be a once-in-a-lifetime opportunity. Shuri Castle will always be my heartfelt example of ichigo ichie—a sentiment that for me also infuses every aspect of this whole sixty-eight-day journey, especially now, in the light of the realizations brought by Covid-19 to the non-poetry reading public about the transient vulnerability of life. As the Romantic English poet John Keats (1795–1821) wrote in his 1816 poem 'Sleep And Poetry':

> Life is but a day;
> A fragile dew drop on its perilous way
> From a tree's summit.

Or, consider the metaphor of life as a fleeting sighting of a passing bird in this poem by Matsuo Bashō (1644–94), Japan's grandmaster of haiku:

> this autumn –
> why am I growing old?
> bird disappearing among the clouds

For me, I am so glad I had not one, but two, bites of this luscious cherry of visiting Shuri Castle, very much against my earlier expectations. The

Phoenix of culture: Residents of Okinawa will especially mourn the loss, from the fire of 2019, of Shuri Castle's main structures and most of the treasures within. However, for sure, rebuilding this cultural wonder will once again demonstrate the resilience of the Japanese people.

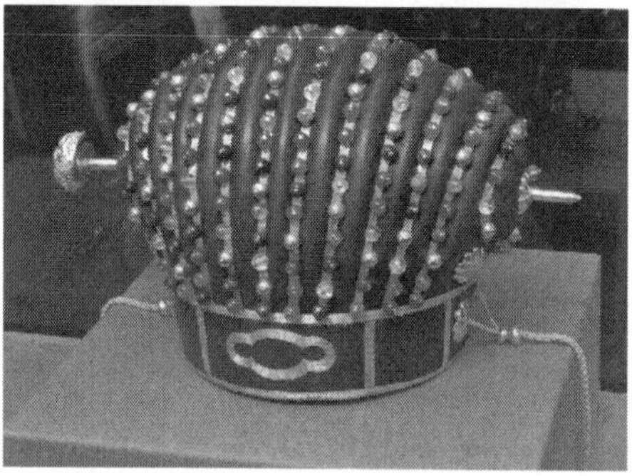

first time I had wanted to visit Okinawa was in the mid-1990s, when, as a twenty-something-year-old journalist at Singapore's main English newspaper *The Straits Times*, I came across an upcoming Asia-Pacific youth conference in Okinawa and wanted very much to attend it. To qualify as a delegate, I had to write an essay on a youth concern to 'compete' for a place. I worked on the piece with gusto, but, as it turned out, was unsuccessful. The regret was a bit hard to bear for the next two decades or so.

Then in 2015, my dream came true. A chance to visit Okinawa emerged when I was invited by the Japan Foundation Asia Centre to visit Japan as a cultural leader of Singapore. I was asked to select up to four destinations. Naturally, Okinawa was on the list, together with Tokyo, Kyoto and Yamanashi prefecture, home of Mount Fuji. Okinawa offered many sightseeing delights, including walking on sand at Kouri Island, views of Shioya Bay from the hills near Mutabaru Village, and sunset at Bise Village. It was most fulfilling to see how much the Okinawan people cherish their home, as captured, for instance, in the name of one bridge in Okinawa—Nirai Kanai, which means: 'a utopia on the other side of the ocean'.

In Okinawa, I had the privilege of also enjoying unique experiences such as making my own lacquer plate at the Naha Traditional Arts and Crafts Centre. Of course, I chose to depict on my plate the brand icons of the two countries: the Merlion and Mount Fuji. Above all, the highlight of this 2015 visit was seeing the majestic Shuri Castle.

My second visit to Okinawa on this sixty-eight-day journey was a precious second shot at ichigo ichie on these islands. This time, one main goal was to pay closer attention to aspects I had missed the first time in 2015. These included some of the exhibition displays inside the castle, in the museum containing a host of fascinating artefacts, from traditional outfits in bright, uniquely Okinawan-style hues to Ryukyuan classical instruments such as the sanshin (literally 'three-stringed', an Okinawan string instrument), as well as getting a deeper sense of the aerial panoramic views over the city of Naha from the hilltop vantage of the castle grounds.

Now, after the fire, amidst the deep pain of loss, what gives hope is that this is no less than the fifth time that Shurijo has been destroyed,

Gift from China: Above the throne in Shuri Castle's main hall (Seiden) is a replica of the seventeenth-century calligraphy penned by Qing dynasty emperor Kangxi and gifted to the Ryukyu king Shou Tei.

following previous incidents in 1453, 1660, 1709 and 1945. So, just like before, this national treasure will definitely rise again, a phoenix of culture whose re-emergence cannot be suppressed. The morning after the blaze, Yoshihide Suga, Japan's Chief Cabinet Secretary, told the media that the Japanese government would do its utmost to help restore the castle. Shuri Castle, scheduled to be a stop on the 2020 Tokyo Olympics torch relay route, would be, after the fire, an even more poignant one, when the Games happen one year after their postponement due to the Covid-19 pandemic.

The castle has a long and rich history, having served as the heart of the ancient Ryukyu Kingdom, which ruled from 1429 to 1879. The three fires that happened during this period were probably due to wood burning, especially during winter. In 1933, it was designated by Tokyo as a national treasure. A fourth fire destroyed the castle in the Battle of Okinawa in the closing days of World War II. Restoration efforts began only in 1989, in preparation for the twentieth anniversary of Okinawa's reversion to Japan in 1972. The reconstruction was completed in 1992, and the castle was designated as a UNESCO World Heritage Site in 2000. With its flaming red facade, sloping tiled roof and Chinese-influenced decorative features, the castle's architecture is distinct from that found on mainland Japan.

Shuri Castle will always be my heartfelt example of ichigo ichie—the awareness that every single experience in life should be cherished and treasured, because it might very well turn out to be a once-in-a-lifetime opportunity.

Okinawa is fascinating for its connections with China and Southeast Asia, and the world outside, for a country and society that is famously protective of its own culture against external influences. Shuri Castle was constructed using both Japanese and Chinese architectural styles, underlining the unique characteristics of the Ryukyu Kingdom, which long served as a transportation hub connecting Japan with China from the time of the Ming dynasty onwards. The Ryukyu Kingdom was a tributary state of China's Ming and Qing dynasties; Ryukyu kings had to be confirmed by the Chinese emperor. A scale model in Shuri Castle depicts such an investiture ceremony by a Qing envoy taking place in the castle's distinctive red-and-white courtyard. The Castle's North Hall, burned down by the 2019 fire, served as the main facility for the Shuri government. The South Hall, also swept away by the flames, was

Lions of Okinawa: The most prominent symbol of centuries-old Chinese influence on Okinawan culture—the shisa lion, seen everywhere including as rooftop gargoyles, as a motif on a kariyushi shirt, and, of course, on Hello Kitty merchandise.

used to receive Japanese Satsuma officials after the Shimazu clan invaded the kingdom in 1609. The destroyed North and South Halls were, therefore, symbolic of the kingdom's concurrent relationships with China and Japan as the halls were used to host ceremonies dedicated to each territory. The Main Hall, which collapsed in front of television cameras during the blaze, was used by the king and his family for administrative and religious purposes.

Okinawa is, in many ways, a very sacred place. Outside Shuri Castle, historical and spiritual awareness has always permeated everyday life throughout Okinawa. Local communities maintain numerous utaki spots, which are places of worship with long histories, and many of them usually go unnoticed by outsiders. The utaki can either be a vast, calming space on the outskirts of a village, or an open, lonely place in a field or forest. Through these places, the community continues to feel the lively presence of both mythological and bloodline ancestors, along with the psychic power that created the islands and their culture.

Okinawa is fascinating for its connections with China and Southeast Asia, and the world outside, for a country and society that is famously protective of its own culture against external influences.

The remotest origins of Shuri Castle are unknown, although historical records suggest its presence from the late fourteenth century. In 1429, three principalities were united into the Ryukyu Kingdom, after which Shuri Castle became the kingdom's palace through two dynasties spanning 450 years. Many Ryukyuan officials were descended from Chinese immigrants. In 1879, Japan annexed the entire Ryukyu archipelago. The Meiji government then established Okinawa Prefecture. The monarchy in Shuri was abolished and the deposed King Shō Tai (1843–1901) was forced to relocate to Tokyo. Hostility against Japan increased in the islands immediately after the annexation, in part because of the systematic attempt by Japan to integrate Ryukyuan culture, including the language, religion and cultural practices.

During World War II, the Japanese 32nd Army decided to station its headquarters in Shuri Castle, fortifying the hill on which the castle stood, and dug a kilometre-long network of military tunnels right beneath

Vistas of paradise: The most famous scenic view of Kabira Bay, as well as other places to visit such as the windswept, rugged coastline of Uganzaki Cape.

it. This decision made the castle a conspicuous target of destruction by the US military. Intense bombardment and gunfire took countless lives, including many civilians, and reduced the castle to rubble and ashes, leaving the American flag on the ruined hill by the end of May 1945.

Since the end of World War II, Okinawa has been a critical strategic location for the US Armed Forces. The island hosts around 26,000 US military personnel, about half of the total complement of the US Forces Japan, spread amongst thirty-two bases and forty-eight training sites. The presence of the US military in Okinawa has caused political controversy on the island and elsewhere in Japan. Japan became a pacifist country with the 1947 constitution, and America was obligated to protect Japan against foreign threats.

During the American military occupation of Japan (1945–52), which followed the Imperial Japanese surrender on 2 September 1945 in Tokyo Bay, the US controlled Okinawa Island and the rest of the Ryukyu Islands. The Amami Islands were returned to Japanese control in 1953. The remaining Ryukyu Islands were returned to Japan in 1972.

Chinese culture and the secrets of longevity

The Chinese elements that have been interwoven into Okinawa's culture include the architectural designs and styles apparent everywhere, as well as motifs such as the dragon and phoenix, essential figures of Chinese mythology. But it is probably the shisa lion that makes its appearance most ubiquitously, including a local modification that makes it uniquely Okinawan.

The shisa is a traditional Ryukyuan cultural artefact and decoration derived from Chinese guardian lions, often seen in similar pairs, resembling a cross between a lion and a dog. In the Chinese universe of symbols, the lions' mouths are usually open, sometimes with a pearl between the lips. In Okinawa, however, one lion (the male) always has its mouth open while the other's (the female) is closed. In 2015, my guide Akiko noted with a laugh that this would probably be thought of by most people as being the opposite of reality, when men are more often considered the less talkative of the two genders. An alternative interpretation could be that the open mouth of the lion perhaps

Woman power: A traditional cultural performance at a restaurant in Naha, with all roles performed by women.

indicates its more assertive and protective nature, compared with the closed mouth of the lioness, suggesting greater submissiveness.

The shisa also features as a motif on the kariyushi shirt, the Okinawan equivalent of the aloha shirt of Hawaii or the bula shirt of Fiji, a cool, casual shirt often printed with motifs featuring leaves and flowers like frangipanis or orchids, fruit like pineapple or birds like parrots. The kariyushi shirt distinguishes itself by featuring unique motifs such as the Okinawan shisa lions. It is worn as smart casual office wear including by business executives and government officials, and also perfectly appropriate for formal occasions.

Another aspect of external influence on Okinawa is reflected in its food. Okinawan cuisine has some distinctive dishes and flavours that

Riding on nostalgia: A bullock cart ride on Taketomi Island transports visitors to bygone eras of rustic simplicity.

set it apart from the washoku cuisine of the mainland. The Okinawan dishes include gōyā chanpurū (bitter melon, or gourd, stir-fried), fu chanpurū (wheat gluten chanpurū) and tonkatsu (tenderized, breaded, fried pork cutlet). The Okinawan dialect word chanpurū even sounds a bit like the Malay word campur (meaning 'mix'), possibly suggesting a borrowed word from the exchange through trade with Southeast Asia, reflecting the mixing of cooking styles, ingredients and tastes from long ago.

Fusion cuisine: Chinese influences on Okinawan cuisine—the goya chanpurū (a salad with bitter gourd, with a name that could have Malay roots) and Okinawan-style noodles.

The generally healthy nature of Okinawan food is part of the explanation for the well-known longevity of Okinawa's residents. Five times as many Okinawans live to the age of 100 compared to their compatriots across the rest of Japan. The Okinawan diet consists mainly of low-fat, low-salt foods, such as whole fruits and vegetables, legumes, tofu and seaweed. Okinawa is categorized as a 'Blue Zone', a term used by American author Dan Buettner for regions of the world where he found that people tended to live much longer than average. Other than Okinawa, the other Blue Zones are Sardinia (Italy), Nicoya (Costa Rica), Icaria (Greece) and amongst the Seventh-day Adventists in Loma Linda, California (US).

What makes Okinawans live so long is their ikigai, the deep sense of purpose in life that makes them eager to wake up each morning to fulfil their passionate pursuits.

In 2002, there were 34.7 centenarians for every 100,000 inhabitants in Okinawa, which was the highest ratio worldwide. Possible explanations are a healthy diet, low-stress lifestyle, caring community, leading active lives, and the deep spirituality of the inhabitants of the island—all of these being ingredients for contentment. We saw glimpses of all these aspects on our visit, reaffirming observations on my earlier visit.

When I visited Okinawa in 2015, I was keen to learn about the Okinawan secrets of longevity. Okinawans would tell you that what makes these islanders live so long is their ikigai, the deep sense of purpose in life that makes them eager to wake up each morning to fulfil their passionate pursuits.

However, I was also told that in recent years, the women of Okinawa are still placed in the top few amongst Japan's forty-seven prefectures for life expectancy, but the life expectancy for men has slipped down the rankings. Most Okinawans I spoke with attributed this trend to recent changes in Okinawan eating habits, pointing a finger at the growing preference for American fast food, an influence from the US military bases on the island. However, I remained sceptical of this theory, given that both men and women eat fast food, so there was possibly some other explanation.

Wonders of nature: Stratocumulus clouds are the icing on this cake of a view of the Okinawan islands seen from a flight from Naha to Fukuoka, while a stunning dawn breaks on Senaga Island.

My own whimsical amateur theory is that stress could be an additional factor, given the limited job opportunities in Okinawa outside of tourism-related sectors and the greater psychological burden on men to bring home the bacon and to appear strong, no matter what happens.

An observation at an airport in Okinawa brought home this point. As often happens in Japan, when a yurukara mascot cartoon character appears, children and women happily go right up to greet the character and to take photos with it, invariably making the two-finger 'peace' gesture and calling out 'cheesu!', as Japanese people like to do whenever they are having their photos taken.

By contrast, the men—especially older men—always stand by the side, hesitant to approach in such situations, worried about what others might think if they were to exhibit any behaviour that might be deemed silly or inappropriate. Bottling emotions up to such an extent cannot be good for emotional and psychological well-being. A few more visits to Okinawa would be necessary to investigate this hypothesis further . . .

Exotic even for the Japanese

Okinawa's many differences from the rest of mainland Japan extend even to the national flower and the favourite national pastime of hanami (flower-viewing). Cherry blossoms (sakura in Japanese) bloom on the island of Okinawa as early as January each year. The temperatures in Okinawa during cherry blossom season range from 16 to 19 degrees Celsius, compared with around the same temperature range in, say, the Yamanashi region in spring from March to May.

While the cultural significance of hanami in Okinawa is similar to that on the mainland, the trees themselves are very different. Unlike the sakura seen in mainland Japan, some cherry trees here originated in southern China and Taiwan. Rather than the pale pink, almost white, blossoms of the Somei-Yoshino trees common to the Kanto and Kansai regions, the Okinawan cherry blossoms, known as Hikanzakura, are a much richer, darker pink. The cherry blossoms also look different, hanging and facing downwards, like pink bells.

Okinawa is exotic even to the Japanese themselves. It is a favourite holiday destination to head for the sun, sand and sea. The captivating romance of the Okinawa islands is brilliantly captured in artworks such as the woodblock series *Eight Views Of The Ryukyu Islands (Ryukyu hakkei)* by the artist Katsushika Hokusai (1760–1849). Okinawa's more than 150 islands offer many opportunities for seaside recreation. The most famous spot is Kabira Bay on Ishigaki Island, which we drove to in our rented car. It is one of only two places in Japan where black pearls are cultivated (the other one being Iriomote Island in Okinawa).

According to legend, one day a ship had anchored in the bay while waiting for the weather to clear up. A small boy took this opportunity to go ashore and explore the nearby area. When he returned, the ship had sailed without him! The boy prayed and prayed for the ship to come back. Then, suddenly a strong north wind started blowing. The bad weather forced the ship to return to the bay, and the boy was able to board it and return to his home in Okinawa. Later in life, the boy, who became a monk, revisited Ishigaki and, in gratitude, built the Kabira Kannon temple.

Okinawa is exotic even to the Japanese themselves . . . The captivating romance of the Okinawa islands is brilliantly captured in artworks such as the woodblock series *Eight Views Of The Ryukyu Islands.*

Aside from Ishigaki, Okinawa's other offshore islands offer many other cultural sightseeing options. The Yaeyama Islands have a rich history and culture going back to before their incorporation into Japan, when they were part of the Ryukyu Kingdom. The Yaeyama Museum on the main Ishigaki Island displays historical and cultural items of these Yaeyama Islands. On display are wooden canoes, pottery, traditional clothing and household ware, as well as items from the various festivals and cultural events held on the islands.

With time to go to only one more island, we chose Taketomi Island, only a ten-minute ferry ride from the Ishigaki ferry terminal. On Taketomi, you can find a beautifully preserved, traditional Ryukyu village, which consists almost entirely of one-storeyed houses covered with red tiled roofs and ample shisa statues to ward off evil spirits, coral stone walls and streets with pure white sand, typical of the urban landscape of Okinawa in ancient times.

Some of the traditional homes in the village are used as restaurants and shops selling local food and crafts. There are no rental cars on Taketomi, with only 300 residents, and aside from taking the bullock cart rides, most visitors either walk or rent a bicycle to travel through the village's white sand roads and to the nearby beaches along the western shore of the island.

The Yaeyama Islands have a rich history and culture going back to before their incorporation into Japan, when they were part of the Ryukyu Kingdom.

For city-dwelling visitors, the most 'unusual' way to get around is by buffalo. Two companies offer water buffalo-drawn cart tours of Taketomi Village. The thirty-minute tours traverse the preserved, traditional streets of the village while a guide talks about the island's history, sings songs and plays Okinawan music on a sanshin. You might notice that the guides rarely offer guidance to the water buffaloes, which seem to know their routes by heart. The buffaloes were once working animals on farms across Japan but have been brought to Taketomi to 'enjoy' a slower pace of life. However, the reality has been somewhat different—it must be a bit of a slog to pull a cart with so many tourists on board.

Fukuoka: Also outward-oriented

On the flight out of Naha, the capital and air hub of Okinawa, the view of the Okinawan islands was spectacular. It was an awesome sight—the shape of the islands, the deep blue of the water merging into the blue of the sky, with the stratocumulus clouds the icing on this cake of a view.

Our next destination was Fukuoka, our other stop in Japan on this journey. Fukuoka was one Japanese city we had never visited, and there were a couple of friends—Steven and his wife Devin—we could meet there. The four of us enjoyed a wonderful chicken hot pot meal together one evening and exchanged views on life in this city that not every visitor to Japan might have on their itinerary.

Touches of Zen: Fukuoka's cultural treasures include the Shofukuji Zen Temple, set amidst serene gardens, and Tochoji Temple, surrounded by office buildings right in the heart of the city.

For instance, not many people realize that Fukuoka has become the largest start-up city in Japan. Fukuoka is not only a happening place for entrepreneurship in technology- and finance-related sectors but is also the business centre of the Kyushu region, with an economy largely focused on services. However, despite its modernity, Fukuoka still retains some old-world charm, like most Japanese cities. This includes Zen temples, some of which, like Tochoji Temple, are now beginning to be a bit hemmed-in by nearby modern office buildings in ever-closer proximity, while others, such as Shofukuji Zen Temple, set amidst serene gardens, have managed to retain some more extensive grounds, which give their heritage structures a grander frontage that has remained unchanged for centuries.

Like Okinawa, Fukuoka also adopts an outward-looking perspective on Japan's neighbouring regions of China and Southeast Asia. For example, a snapshot of this can be found at the Fukuoka Asian Art Museum. The museum is known for its progressive strategy for interaction with different Asian cultures. This focus on exchange with continental Asian culture shows an orientation to the region and the outside world that reflects Fukuoka's history as a gateway to the rest of Asia since ancient times.

Like Okinawa, Fukuoka also adopts an outward-looking perspective on Japan's neighbouring regions of China and Southeast Asia.

There are at least two oil-on-canvas works on display by Singaporean pioneer artists. Liu Kang's (1911–2004) work *Old Street* (1950) depicts a shophouse that could have been in Singapore or Malaysia. In the painting, three of the shophouses are boarded up on the ground floor, suggesting that these occupants have not yet recovered from the destruction of World War II. Georgette Chen's (1906–92) *Cityscape Of Beijing* is a view of China's capital that must have carried much personal meaning for her as the annotation informs the viewer that the artist kept an identical painting of this view with her till her death.

The interface with Southeast Asia that one finds in Okinawa and Fukuoka typifies the more global mindset of modern Japan. To some foreigners, this might seem like a relatively new brand attribute, but in

Singapore in Fukuoka: *Old Street* (1950) by Liu Kang and *Cityscape Of Beijing* (1950) by Georgette Chen at the Asian Art Museum.

Cultural ballast: Just as heritage is preserved as with this doll's distinctive Okinawan yukata at Yaeyama Museum, a resilient Japan should surface strongly from this pandemic.

fact, Japan has been investing a lot into promoting this aspect of itself for some time. The modern, cosmopolitan Japanese who is articulate, well-read, well-informed and deeply interested in connecting with other cultures is reflected in several of my very good Japanese friends, including Misako, whom we met for coffee on a brief stopover at Haneda airport in Tokyo on our flight in from Mexico in transit to Okinawa.

Meanwhile, many other expressions of Japanese soft power are active. The 2020 Tokyo Olympics, when they eventually happen, are set to be a game-changer for international perceptions of Japan. Looking out over the runways at Haneda airport, a few parked planes emblazoned with Tokyo 2020 graphics were a visual anticipation of this next major nation branding event for Japan. Parts of Japan, such as Okinawa, which are at the edge of Southeast Asia are, in a sense, at the interface or forefront of advancing brand Japan. As the effects of the Covid-19 pandemic recede and more 'travel bubbles' are established with other countries in Asia, these interfaces with Southeast Asia might play more important roles in restoring connectivity between Japan and the rest of the world. As for travellers from countries further away, taking time to appreciate places such as Okinawa and Fukuoka, with their many ancient and modern attractions, will continue to give a deeper sense of these new faces of the land of the rising sun looking to a post-pandemic dawn, just as Shuri Castle will surely shine once again as a beautiful interface with the rest of the world.

Taiwan

Spirit of enterprise

The haiku poem and ink sketch in this haiga artwork position the Taipei 101 skyscraper as the towering brand icon of Taiwan and a symbol of the territory's quest to forge its own destiny in an uncertain world. From this emerge its place brand attributes of spirit and enterprise, as it keeps rising above the weight of nostalgia for the motherland, across the channel of its past, towards a challenging future that will require drawing upon all its reserves.

TAIWAN
Standing tall
from nostalgia here's
the future anew, even
with the strait so near
許木松
2020

Deep reserves: Taiwan has had to draw deeply upon its resources of enterprise to find its own way in the world.

A symmetry of skyscrapers: Taipei and Dubai

And so, we reached the last stop before home on this sixty-eight-day journey around the world. From here, it would be just one small skip and hop back to Singapore. Taipei was a stop that suited the flight schedule, but there were also other elements that made it a fitting final destination. Most striking of all was a symmetry of skyscrapers, aligning the Taipei 101 tower with the Burj Khalifa, iconic symbol of our first destination, Dubai, where we had started this voyage.

Taipei 101 has conquered the world's imagination as the most photographed and recognizable landmark of Taiwan, eclipsing, with its height and commanding skyline presence, its predecessors such as the Chiang Kai-Shek Memorial Hall, which is a giant of cultural and historical heritage but a midget by comparison based on physical stature. The 101-floor Taipei 101 claimed the title of world's tallest building when it opened in 2004, which it held for six years before relinquishing it to the 160-storey Burj Khalifa in Dubai.

Taipei 101 plays a key role in the territory's internal brand-building. There is nothing like an edifice that reshapes the skyline to draw all eyes to everything that it connotes.

Formerly known as the Taipei World Financial Center, Taipei 101 measures 509 metres from ground to top, making it the first skyscraper in the world to break the half-kilometre mark in height. It was built to last, to withstand typhoon winds and earthquake tremors. Its design incorporates many engineering innovations and has won numerous international awards. Upon its completion, Taipei 101's height exceeded that of the Petronas Towers in Kuala Lumpur, Malaysia, which were previously the tallest inhabited skyscrapers at 451.9 metres.

Taipei 101, the current tenth tallest skyscraper in 2021, holds the certification as the world's largest 'green building' for its environmentally friendly features such as its roof and façade water recycling system, which meets almost a third of the building's water needs. Its shopping mall and its indoor and outdoor observatories draw visitors from all over the globe. Taipei 101's New Year's Eve fireworks display is a highlight

New heights: Taipei 101—symbol of Taiwan's traditional culture and most modern aspirations.

for residents and all Taiwanese citizens watching on TV and is also a regular feature of international TV broadcasts with their roundup of New Year's Eve celebrations around the world.

Like other such skyscrapers, Taipei 101 plays a key role in the territory's internal brand-building. There is nothing like an edifice that reshapes the skyline to draw all eyes to everything that it connotes. The tower's design and aesthetic features—its visible look and feel—certainly help foster national and cultural identity, belonging and pride. These more concrete facets feature alongside the abstract ideas evoked, which revolve around all the inspirations that can be derived from what the structure represents, including ample resources, long-term vision and planning, creativity, engineering excellence, ambition and promise for the future.

What makes Taipei 101 an even more fitting city and country symbol are its universal, culturally meaningful features, which are richer in this tower than in most other towers elsewhere. It is appropriate for its acknowledgement of the past as well as its affinity with the future. Universally, the significance of the numbers associated with the building is remarkably rich. The height of 101 floors signals all the new years that are to follow (1 January = 1-01). It symbolizes high ideals by going one better than 100, a traditional number of perfection. The number also evokes the binary numeral system used in digital technology, thereby aligning with digitization, the defining aspect of human development in this twenty-first century and third millennium, as well as the fourth industrial revolution.

What makes Taipei 101 an even more fitting city and country symbol are its universal, culturally meaningful features, which are richer in this tower than in most other towers elsewhere.

Culturally, Chinese symbolism is represented throughout. The main tower features a series of eight segments of eight floors each. In Chinese-speaking cultures, the number eight is associated with abundance, prosperity and good fortune. The repeated segments simultaneously recall the rhythms of an Asian pagoda (a tower linking earth and sky), a stalk of bamboo (an icon of learning and growth) and a stack of ancient Chinese ingots or money boxes (a symbol of abundance).

There are even more cultural aspects. In jest, some people have likened the building's shape to a stack of boxes as used for Chinese takeaway food. The four discs mounted on each face of the building where the pedestal meets the tower represent coins. The emblem placed over entrances shows three gold coins of ancient design, with central holes shaped to

Eye over Taipei: Views from the eighty-ninth floor observatory of Taipei 101 tower, including sights of buildings such as the 'twisting' Agora Tower.

imply the Arabic numerals 1-0-1. The whole structure incorporates many shapes of squares and circles to reach a balance between yin and yang.

Curled figures called ruyi appear throughout the structure as a design motif. Although the shape of each ruyi at Taipei 101 is traditional, its rendering in industrial metal is plainly modern. The ruyi is a talisman of ancient origin associated in art with heavenly clouds and connotes healing, protection and fulfilment. It appears in celebrations of the attainment of new career heights. The sweeping curved roof of the adjoining mall culminates in a colossal ruyi that provides shade for the pedestrians below. What's more, at night, the bright yellow gleam from its pinnacle casts Taipei 101 in the role of a symbolic candle or torch upholding the ideals of liberty and welcome. From 6 p.m. to 10 p.m. each evening, the tower's lights display one of the seven colours of the spectrum.

A place in the region

Somehow, somewhere, human histories are all connected, and the history of Taiwan is closely intertwined with that of our previous stop on this journey: Okinawa. The Yuan Dynasty Chinese traveller Wang Dayuan used 'Liúqiú' as a name for the island of Taiwan in his book *Daoyi Zhilüe*, written circa 1339, translated as *A Brief Account Of Island Barbarians.* The place where he landed was closest to Penghu, or the Pescadores Islands, an archipelago of ninety islands southwest of Taiwan. The name Liúqiú was used for the Ryukyu Islands in general or Okinawa, the largest of them. The name Ryūkyū is the Japanese form of Liúqiú. The name also appears in the *Book Of Sui*, the official dynastic history of China's Sui Dynasty (581–618).

Taiwan's connections with the region stretch even further back. This land was joined to the Chinese mainland in the Late Pleistocene era, until sea levels rose about 10,000 years ago. Fragmentary human remains dated 20,000 to 30,000 years ago have been found on the island, as well as later artefacts of a Palaeolithic culture. Around 6,000 years ago, Taiwan was settled by farmers, most likely from mainland China. They are believed to be the ancestors of today's Taiwanese aborigines. The languages they spoke belong to the Austronesian language family but show much greater diversity than the rest of the family. This network of linguistic communities spans a huge area, from western Southeast Asia to Madagascar (making another intriguing link to another of our earlier destinations on this journey) and, to the east, as far as New Zealand, Hawaii and Easter Island. This has led linguists to propose Taiwan as the Urheimat ('original homeland' in historical linguistics) of this language family, from which seafaring peoples dispersed across Southeast Asia and the Pacific and Indian oceans.

Taiwan's quest for independence remains a dream tinged with anguish—this perhaps is its defining place brand attribute.

In the seventeenth century, Dutch rule opened the island to mass Han immigration. After a brief period of rule by the Kingdom of Tungning, which held sway over parts of the southern and western areas of the island, the island was annexed in 1683 by the Qing dynasty of China and ceded to the Empire of Japan in 1895. Following the World War II surrender of Japan in 1945, the new Republic of China, which had overthrown and succeeded the Qing dynasty in 1911, took control of Taiwan on behalf of the World War II allies. The resumption of the Chinese Civil War led to the loss of the mainland to the Communist Party of China and the flight of nationalists and their followers to Taiwan in 1949. Although the Taiwanese government continued to claim to be the legitimate representative of China, since 1950 its effective jurisdiction has been limited to Taiwan and numerous smaller islands.

Like many other territories, Taiwan retains aspects of the legacy of imperialism. This is reflected in the various names for the island of Taiwan that remain in use today, each derived from explorers or rulers during a particular historical period. The name Formosa dates back to around 1542–44, when Portuguese sailors sighted an uncharted island and noted it on their maps as 'Ilha Formosa' ('beautiful island'). But here, the lingering influence of Western expansionism is slight, compared with many other places.

Till today, there is a particular poignancy about Taiwan's state of national identity. A full sense of self-determination has always

Scale and symbolism: A traditional Chinese painting in a hotel lobby shows a lone human figure contemplating man's insignificance in nature, in contrast to a graphic rendition of the Taipei landscape dominated by the Taipei 101 tower, which depicts a different power relationship between man and nature.

eluded it, in contrast to the kind of untrammelled sovereignty that is safeguarded on a daily basis in other places, such as Singapore. Whereas Singapore had full self-government thrust upon it with separation from Malaysia in 1965, Taiwan's quest for independence remains a dream tinged with anguish—this perhaps is its defining place brand attribute.

Taiwan's situation is captured well in the poem 'Nostalgia' by the prominent Taiwanese poet, essayist and translator Yu Guangzhong (1928–2017), who lectured at the National Sun Yat-sen University in Taiwan for thirty-two years. This poem came to symbolize the aching separation, displacement and longing for cultural unity felt by many in mainland China and across the Chinese diaspora. The poem, which has been included in the standard literature curriculum in China and Taiwan and is taught to generations of students, subtly but powerfully draws on imagery familiar to a generation of homesick Chinese. Even today, many Taiwanese—young and old—can recite the poem in its entirety:

Nostalgia

When I was a child,
Nostalgia was a tiny postage stamp,
I, on this side,
My mother, on the other.
When I was older,
Nostalgia became a ship ticket,
I, on this side,
My bride, on the other.
Later,
Nostalgia was a squat tomb,
I, outside.
My mother, inside.
And now,
Nostalgia is a coastline, a shallow strait.
I, on this side,
The mainland, on the other.

Yu's poem sums up a lifetime of longing, evoking imagery that encapsulates the whole bundle of emotions wrapped around being kept apart from something that has deep bonds of kinship. Words such as 'shallow strait' drive home the poignancy of separation, a deep sense of 'so near, yet so far'. Known for drawing from Chinese culture with a Western aesthetic, Yu had a passion for Western literature and in 1957 published his Chinese translations of works including Ernest Hemingway's novella *The Old Man And The Sea*. The Hemingway connection marks another link to an earlier destination on this trip—Cuba, where Hemingway lived for many years. The deepest similarity between Taiwan and Cuba is that both places are islands adrift in seas of political turmoil.

Yu was a college student in mainland China when his family joined the wave of nationalists who fled to Taiwan after they were defeated by Mao Zedong's communists in the Chinese Civil War in 1949. Like many exiled Chinese who had left behind relatives, friends and homes,

Complicated nostalgia: Sunset view from Taipei 101, looking to the west, in the direction of the 'motherland', China.

he could never quite shake off his yearning for the 'motherland'. In his 1966 poem 'Percussion', he wrote:

> The Yellow River flows torrential in my veins
> China is me I am China

Though Taiwan and mainland China are separated only by a narrow strait, travel between the two territories was largely prohibited after 1949. By 1971, Yu said he was tormented by the uncertainty of when, if ever, he would be able to return to his ancestral homeland. 'It was under that kind of pressure that I wrote "Nostalgia",' he told Phoenix TV, a Hong Kong broadcaster, in 2014. 'It only took me 20 minutes to write because the feeling had been in my heart for over 20 years.'

During a visit to the US in 2003, the Chinese premier at the time, Wen Jiabao, quoted Yu's poem 'Nostalgia' in a discussion to make the case for an eventual reunification with Taiwan, which the Communist Party considers part of China, a 'renegade province' as it is sometimes referred to in the Western media. 'The shallow strait,' Wen said, using the Chinese phrasing from Yu's poem, 'is our greatest nostalgia and biggest sadness.'

Duty to a larger cause

For me, Taiwan evokes memories drawn from my brief visits to the southern region twice for national service military training in the late 1980s. A poem in my first verse collection *A Brief History Of Toa Payoh And Other Poems* (published in 1992) is an attempt to recreate a scene set in Taiwan, recalling an experience of deprivation from the creature comforts of everyday life in Singapore. The poem, titled 'Memories Of A Barren Plain', describes how soldiers, tired and hungry, yearning for rest and a midnight snack, must still press on with military training out of a sense of duty and service. The poem opens with an imagined mental picture of what life must be like for the farmers of Taiwan, who also have to plough the earth, and for whom back-breaking labour is a daily routine:

life is always hard for farmers
surreptitious winds
cut to the very bone
even in April,
and though buffalos roam,
the plains are dry as hay
and thorny thickets thrive

In similar fashion, soldiers on a defence exercise must also operate in the same environment exposed to the elements. They also rely on farming implements to dig a different type of trench. Hence, while the words 'life is always hard' are the same, they apply differently to the two groups:

but the vigil must be kept
for any light or sound
on the drowsy horizon
of duty and honour
life is always hard for soldiers

Here in this snippet of the hardship of military service is perhaps just a fraction of what the most patriotic Taiwanese must feel all the time. But what connects the two scenarios of farming and military service is what keeps one going, a commitment to a cause larger than oneself, which is a common thread between Taiwanese and Singaporeans.

About a decade later, I had another poem set in Taiwan published, this time in my third verse collection *The Ocean Of Ambition* (published in 2003). This satirical poem, titled *A Brief History Of Performance Appraisal*, draws observations from a visit to the National Palace Museum in Taipei on a vacation in 2002, my third visit to Taiwan. Many such museum exhibitions are collections of extravagant gifts to monarchs of the past, to win the favour of the palace:

in days of China's yore –
as these showcases remind –
loyal courtiers and eunuchs,
flurrying with flapping sleeves,

State dinner: A great idea that more restaurants should also adopt—offering diners a wide selection of national flags that they can place on their tables. As bonds of international friendship are forged over a good meal, the flags provide an evocative table centrepiece.

vied and tripped to serve up
on the platter of obeisance
the choicest curio cabinets
crammed with scores of delicate delights

Very soon, a culture of patronage forms, such that it becomes impossible to avoid being sucked into this kind of system of 'performance appraisal', which is still observable in the corporate world today:

today, in glass offices
of segmented systems
just as layered, even if they also flap,
officers of rank, slotted into
travellators of forbidden corridors,
procure items of a different curiosity

This is duty and service of a different kind from that on the hill slopes described in the earlier poem. The self-sufficiency and self-reliance that are key brand attributes of the Taiwanese people place in stark contrast the very different organizational culture of ancient times in the 'motherland'.

To be sure, the National Palace Museum is one of Taiwan's greatest attractions. It houses more than 650,000 pieces of Chinese bronze, jade, calligraphy, painting and porcelain and is considered one of the greatest collections of Chinese art and objects in the world. The Kuomintang—Taiwan's Nationalist Party formed in 1919, which was the sole ruling party from 1928 to 2000—moved this collection from the Forbidden City in Beijing in 1933, and part of the collection was eventually transported to Taiwan during the Chinese Civil War. The collection, estimated to be one-tenth of China's cultural treasures, is so extensive that only one per cent is on display at any time.

China has said that the collection was stolen and has called for its return, but Taiwan has long defended its control of the collection as a necessary act to protect the pieces from destruction, especially during the Cultural Revolution between 1966 and 1976, when anything remotely bourgeois was destroyed. Relations regarding this treasure have warmed in recent years; Beijing Palace Museum Curator Zheng Xinmiao has

First and last: A Qiblah sign (indicating the direction that a Muslim should face in prayer) at the Taipei 101 tower makes a topographical connection between the first stop (Dubai, in the general direction of the Middle East) and this last stop on this sixty-eight-day journey.

said that artefacts in both Chinese and Taiwanese museums are 'China's cultural heritage jointly owned by people across the Taiwan Strait'.

Despite its geopolitical uncertainties, Taiwan actually has a lot going for it as an economy and society. With 23.7 million inhabitants, it is the most populous state and largest economy that is not a member of the United Nations. Taiwan's export-oriented industrial economy is the twenty-first largest in the world, with major contributions from steel, machinery, electronics and chemicals manufacturing. One of the four famous 'Asian tiger' economies of the 1990s, Taiwan, by all measures, has all the qualities of a developed country, ranking fifteenth in GDP per capita. It is also highly placed in terms of political and civil liberties, education, healthcare and human development. Its management of this pandemic has also won international praise.

Another reason we had wanted to visit Taiwan on this journey, my fourth visit, was to meet our friends Gina and Peter in their hometown. They very kindly hosted us to a sumptuous dinner at a Taiwanese restaurant on one of the top floors of Taipei 101. Their own personal and professional success reflects the spirit of enterprise that is evident in many Taiwanese people. Their international outlook and open mindset are just two of their other admirable qualities.

National history has influenced the Taiwanese people's sense of the world and of the extra effort needed to invest in their own futures to seek a better life for themselves and their families. This they have done with admirable fortitude.

To a large extent, the national history of the Taiwanese has been a story of challenges, uncertainty, external pressures and disappointments in relations with other nations. This must have influenced the Taiwanese people's sense of the world and of the extra effort needed to invest in their own futures to seek a better life for themselves and their families. This they have done with admirable fortitude. Looking out of the window right beside our table, the twinkling lights all around us are reminders of the many lives seeking to be aligned towards a similar course in this shared future.

At this dinner at the Taipei 101 tower, I was reminded once again of the Japanese concept of ichigo ichie, to cherish every experience by appreciating coincidences that make that encounter even more significant. During our visit to the observatory earlier that day, we saw on one windowsill a Qiblah sign (indicating the direction towards Mecca, which a Muslim must face when praying), something one sees in some hotel rooms, but this was the first time I had seen this item in a skyscraper observatory.

The Qiblah sign makes a topographical connection between our first stop on this sixty-eight-day journey (Dubai) and our last destination (Taipei). It reaffirmed the symmetry of starting and ending our journey with a visit to a major skyscraper, providing an iconic vantage point at each end of the trip. Thus, we could get a bird's eye perspective of the future to begin with, and a broad scope to look back and reflect on how far we had journeyed, how much we had seen and absorbed, and how much more there was for us to uncover ahead.

And so, in Taipei, we found ourselves in a city reaching out—and up, literally—as far as it can towards modernity and the future, to grasp all the self-assurance and hope that the skies can yield, just as we had when we started out on this journey in Dubai. The human spirit has depths of fortitude, and humanity holds within itself as much vigour as vision. There is so much of the world to see and experience, so much energy and enterprise to admire, and, most of all, so much more aspiration to feel affinity for.

Farewell view: This photo, taken in a taxi on the move, captures one last sight of Taipei 101 and the Taipei skyline on the way to Taoyuan International Airport, for the final flight home moving into day sixty-eight.

Epilogue

This sixty-eight-day trip was indeed a voyage of discovery and learning. And there was so much to contemplate and cherish.

My wife Dora and I started our journey around the world with a bird's eye view from the top of the Burj Khalifa tower in Dubai, in the United Arab Emirates. There, we tried to peer towards the desert horizon and, despite the dim light of dawn, with the shroud of haze from the sand and dust reaffirming the expanse of the world, we could see a lot in a broad sweep. And yet, because we were so high up, we could make out only so much detail. This was a reminder that—just as it is in life—there is a lot that can be revealed only by making the effort and opening the mind to go down to be at one with the ground.

In Oman, in the halls of the Royal Opera House in Muscat, we saw how ancient tradition can not only coexist with modernity but, in fact, can also rejuvenate itself with a renewal of perspective and broadening of mindset and outlook. This develops best when inspired from the top and participated in from bottom up—and never giving up, like the turtles that clamber ashore year after year.

Madagascar provided a case study of the psyche of one of the most fascinating of island nations. It was heart-warming to think about the intriguing origins of the roots of international kinship through the first ancient migrants from Southeast Asia, long before the era of French colonization. It was mind-boggling to consider the courageous routes taken to seek their fortunes in flimsy canoes across treacherous oceans. Here, similar energy and enterprise are exemplified by the territory's unique wildlife, for whom the bounty of nature also holds rich rewards, from the steady stature of zebus to the daring dexterity of lemurs.

Kenya is a major country brand with immense resources, including its national parks with their awesome wildlife, as well as its outstanding British colonial heritage. It was absorbing to see how even more brand value could be derived from other aspects such as its traditional Maasai people, ensconced deep in the savannah, yet grappling with the encroachments of modernity, more than touched by globalization through information, imports and tourism.

Spain should, by rights, wield stronger soft power globally, given its inherited resources from the scale and legacy of the Spanish empire. Instead, putting aside the mantle of imperialism, a city such as Bilbao stepped forward into global influence to become a model for place brand-building for the future with its 'Guggenheim effect' of urban, economic and social transformation.

Colombia demonstrates the power of culture through its superstar artists such as Botero, showing how the arts can produce more effective brand ambassadors than anyone else from any other sector. The key is how artists are embraced and celebrated in their homeland, so as to magnify even more their international appeal and reinvigorating power for the country.

Costa Rica's country brand X factor is its enlightened stewardship of nature, sustaining the endowments of the wild with human respect and care. The sumptuous wildlife and sophisticated conservation offer so much to visitors, and these qualities of natural vigour are also aligned with the people, increasingly embraced and exhibited by them too, as they move together towards the future.

Cuba is an island of wonder, so full of rich culture and history, coveted and contested by external forces for so long. It makes for such a potent brew of so many culture clashes, especially at the interface between communism and capitalism, and between the Spanish- and English-speaking worlds. Come what may, the people's exuberance, encapsulated by the rhythms of salsa and the instinct for liberty, are an enduring triumph of the human spirit.

Panama is not an island, but feels like one, given the distinctiveness of its geography and place in the world. Indeed, it is actually quite the opposite—not self-contained like an island but opening out like its iconic Canal that links two universes across two oceans. Like a quintessential ambassador of globalization, in an era of nascent deglobalization, its modern vibe derives from its natural bend towards cosmopolitanism and its invigorating orientation to rejuvenation and reinvention.

Mexico has a long and full history, something that it shares with other places that have also hosted major ancient civilizations. As exemplified by the contrast between Teotihuacán and Mexico City, the struggles of hierarchies and societies of eras past provide a poignant backdrop to centuries of sustained human strife, burdens underlying more recent advances towards a higher quality of life.

Guatemala, with its majestic seats of history like Antigua, offers sights that have survived and still speak eloquently of centuries of colonial influence. This is what gives a people the cultural and spiritual sustenance to contend with challenges arising from the social and other concerns of today.

Japan is a place of resilience and rebuilding. Okinawa, at the edge of Southeast Asia, is an emblem of centuries of contending with foreign powers while striving to retain the core of a unique identity of cultural fusion, exemplified by its distinctive shisa lions. Shuri Castle, destroyed so many times, continues to inspire with its call for unquestioned fortitude as it re-emerges, once again, like a phoenix.

And finally, in Taiwan, over dinner at the top of Taipei 101 tower, we appreciated the symmetry of wrapping up our global circumnavigation with a skyscraper overview, just as we had started with in Dubai. We were also inspired, once more, with the spirit of a territory that has always pulled itself up from the trauma and angst of history to chart grand new paths for itself.

And so, a note about the future. This book opened with a prologue of musings infused with the spirit of the Japanese concept of ichigo ichie, applying a once-in-a-lifetime consciousness to the very idea of living, and to travelling, for sure. This ancient wisdom has certainly been borne out amidst the wreckage wrought by Covid-19, looking back on this journey around the world in sixty-eight days. In a world so huge, there is so much more to heed, and to hold in our hearts.

The Author

In a guayabera shirt, on the way to Buena Vista Social Club in Havana, Cuba.